WIRELESS INTERNET CRASH COURSE

ROMAN KIKTA

AL FISHER

MICHAEL P. COURTNEY

McGraw-Hill

New York • Chicago • San Francisco • Lisbon
London • Madrid • Mexico City • Milan • New Delhi
San Juan • Seoul • Singapore • Sydney • Toronto

McGraw-Hill

A Division of The McGraw·Hill Companies

1 2 3 4 5 6 7 8 9 0 DOC/DOC 0 9 8 7 6 5 4 3 2 1

ISBN 0-07-138212-7

The sponsoring editor for this book was Stephen S. Chapman, the editing supervisor was Steven Melvin, and the production supervisor was Sherri Souffrance. It was set in Fairfield by Patricia Wallenburg.

Printed and bound by R. R. Donnelley & Sons Company.

McGraw-Hill books are available at special quantity discounts to use as premiums and sales promotions, or for use in corporate training programs. For more information, please write to the Director of Special Sales, Professional Publishing, McGraw-Hill, Two Penn Plaza, New York, NY 10121-2298. Or contact your local bookstore.

 This book is printed on recycled, acid-free paper containing a minimum of 50 percent recycled, de-inked fiber.

To my wife, Jennifer, an ardent Internet and wireless user—for all your love and support.

To my 16-month-old daughter, Katherine Grace, who's already fascinated by handheld devices and will grow up knowing only a "wireless world."

To my parents, who don't understand the technology but nonetheless are captivated by it all—for your encouragement.

Roman

Dedicated to my loving wife, Peggy, and all our children David, Patrick, Amy, and Zachary. Their love, support and understanding made this book possible for without it, I could not have persevered the long nights and weekends of writing instead of spending time with the family.

In memory of my late Mother and Father who taught me to set goals and strive to achieve them; and above all else, to hold an education in the highest esteem. They instilled in me that you could always learn something new.

Al

Dedicated to my parents, Earl and Colette Courtney, who, in addition to teaching me the value of hard work, have helped me to see that not everyone is an early adopter, and have served (albeit unknowingly) as my faithful indicator of when a particular technology was finally ready for the non-technical mainstream. Through their cautious but curious approach in adopting digital technologies, I am constantly reminded of the need for day-to-day functionality, ease of use, and above all, value for money.

Mike

CONTENTS

FOREWORD

Many people worldwide experience Internet connectivity on their computers at work, enjoying a high-speed, seamless connection through corporate networking. However, less than 10–15% of these same people enjoy a broadband connection for personal Internet access at home; the number of broadband mobile users is significantly less than that at this point. Many of these individuals who delight in high-speed access at work are disappointed with their home experience and highly disappointed with their mobile access experience. This results in a limitation of personal use or SOHO work objectives. Today, the primary options available to fixed broadband customers are cable, DSL, satellite, or point-to-point and point-to-multipoint wireless. These are good technologies but insufficient to address all global broadband needs due to range, flexibility, or physical connection requirements. In many cases, there is a virtual monopoly of these existing broadband solutions that makes them expensive and potentially unattractive to the end user. Mobile users have even fewer choices for broadband access but the marketplace is changing. Competition and market penetration require new technology for wireless mobile and fixed broadband access systems. This is the dawn of the Wireless Internet. The team here at Navini Networks believes that narrowband, slow-speed connections will be things of the past in a few years and that wireless solutions must be a significant part of that change.

The world of wired telecommunications has developed for well over 100 years and continues to advance daily at an astonishing pace. At the same time user behavior is also evolving and untethered access is considered an increasingly important requirement. Although wireless telecommunications has existed for the last 30–50 years, it only became popular with the advent of mobile cellular communications in the last 15 years. Today, the majority of voice calls are divided between cordless phones and cellular phones, skipping the basic wired phones of the traditional wired world of telecommunications. The transfer over to wireless options has been especially acute in the last 5 years as the price/performance of wireless technologies makes them an acceptable alternative to wired technologies.

The rush to wireless in the last mile has been dominated by voice, with multimedia data services in most cases restricted to niche opportunities due to cost and throughput (bits per second). However, the utility of nomadic and mobile access that wired solutions could not provide is intrinsic to all uses of wireless solutions including multimedia. At the same time it is highly unlikely that the end user will agree to pay more than a fair increment in price for that benefit while at the same time tolerating significantly lower throughput compared to wired options. However, it appears, that wireless technologies are at the point now where they will be able to deliver the price/performance to meet an ever-increasing number of users' multimedia needs. A wired broadband Internet user typically stays online 2 to 3 times longer than a dial-up user. Imagine a further multiplier of use if the wires are removed and data speeds are still significantly greater than dial-up! This paradigm is what can be characterized as the Wireless Internet, a world of always on, always connected, and untethered by wires.

In mid-2001 we are provided glimpses of the possibilities of what the Wireless Internet could be with services like Docomo's i-mode and Sprint's Wireless Web, but these are just the early covered wagon days. Now imagine as throughputs over wireless migrate from 14.4kbps to 10 times that with mobility and then 100+ times that with other Wireless Internet technologies like 802.11 and MMDS Generation 2 systems. You are now imagining the equivalent of the space age.

A key observation regarding enabling technologies that is highly useful is that you usually cannot anticipate all the potential uses; basically, we do not possess a crystal ball to predict the future. Therefore, the killer apps of the Wireless Internet will either evolve or change from what the predictions anticipated. Wireless broadband access and all forms of the Wireless Internet will produce significant changes in the way we interact with the Internet and even each other. It will generate massive changes in the multimedia experience for all of us. The authors of this book provide explanation and background to explain the intricacies of the Wireless Internet and they also provide some glimpses of what the Wireless Internet future holds for all of us. At Navini, we believe in the future of the Wireless Internet and I am excited to have the opportunity to write the foreword to this book.

The Wireless Internet is at the start of its journey and will not be composed of one technology or solution. The vision will be an interworking experience of numerous fast, economical wireless solutions that will allow everything from personal super short connections of a few feet, to LAN connectivity, to WAN last mile service and finally nomadic and mobile high-speed connections. The access method of choice may soon be without wires via the Wireless Internet and feed into the large optical backbones allowing connectivity by anyone when and where it is needed as opposed to only where an RJ45 jack exists.

Alastair Westgarth
Chief Executive Officer
Navini Networks
August 2001

PREFACE

The wireless industry is on the threshold of a fascinating era—a convergence of wireless communications and the Internet that enables connectivity from anywhere. Connectivity from anywhere is the truly unique value-added component that wireless brings to the equation. The wireless telecommunication industry is experiencing record growth. Over 860 million mobile phones are in use worldwide and this number is growing at an average rate of 18 percent per year. We believe the Wireless Internet will be a key driver of industry growth for the next decade. As new technologies, applications, and content abound, even seasoned industry veterans find it difficult to keep abreast of the latest developments. Many new protocols have emerged such as Wireless Application Protocol (WAP), Compact HTML (c-HTML), i-Mode, and J2ME just to name a few. Each of these protocols is aimed to merge the Internet and wireless communications into a tool portfolio of Internet access for the mobile user. More to the point, many of these protocols proclaim to be the de facto standard for accessing Internet content from a wireless device. Not a single day passes without multiple articles about one or the other "standard" surfacing in various wireless trade publications. Many experts are very positive about the particular protocol in question and many others are quick to point out its shortcomings.

This book was written as a tutorial on the many key issues and opportunities relating to the Wireless Internet, pro and con. It is not meant to be a conclusive reference for software developers for any one protocol. There are many excellent books on "how to" for c-HTML, WAP, or one of the many other protocols. (See Appendix A for a list of reference publications.) *Wireless Internet Crash Course* is written for wireless telecom managers, developers, network managers, engineers, technicians, sales and marketing personnel, investors, and entrepreneurs. We hope to not only impart information but hopefully, to inspire the more innovative to generate an idea or two. Many of you will find some chapters of more interest than others; some will probably skip certain chapters, to the disappointment of the authors, but that's why we included so many different topics.

Each chapter presents a topic in a thorough, yet not overly complicated manner. Reading standards documents can be very challenging: we have participated in many standards meetings and witnessed first hand, the nodding heads, the drooping eyelids, and the disruptive noises of engineers as they doze. Fortunately, they do wake and complete the standards because we have the proof in the finished documents!

We want this book to be readable without overburdening the reader. It is meant to be as complete and up-to-date as possible. In Appendix B are URL references to many interesting sites containing developer tool kits, complete standards documents, sources of Wireless Internet devices, and some other darned interesting information. The material presented in this book makes this a virtual, single source tool kit for information on the Wireless Internet. Visit these Web sites often because they are constantly updated with the latest information available about the Wireless Internet. For your convenience, we have also established a Web site with links to the sites listed in Appendix B at www.genesiscampus.com/appendixb.htm so that you may let your mouse do the work.

Chapter 1 is an introduction to the Wireless Internet. No book on this subject could be complete without a brief history of communications and the Internet, so we take an historical look at how it all started, why it was started, and some basic details on how it works. Any discussion on the Wireless Internet must also mention some of the participating companies that are the driving force behind it. (Actually, we'd gladly mention all companies but therein lies a big problem. Between the time that we started this book and the time it appears on your bookseller's shelves, many new or existing companies will enter and probably just about as many will cease to participate in this dynamic market!)

We define what the Wireless Internet is and perhaps what is it not. Trends toward an increasingly mobile society using wireless communications are making world standards bodies work feverishly to implement Third Generation Digital Cellular Systems. Part of the driving force behind this standards work is the Wireless Internet; requirements for implementing mobile e-commerce sites will change as new 3G systems replace the current 2G systems, and an outline of these requirements, current and future, is included in this chapter.

The transition from a circuit-switched architecture to a packet-based network requires new technology. The demands on current networks require many new components. We will discuss Internet Protocol (IP) and the transition to a world of Voice-over-IP (VoIP) and an all-IP network. New products such as media gateways and soft switches are emerging as the workhorses of the next millennium. These new switch technologies not only result in less space and lower power requirements but they allow distributed processing utilizing protocols such as SIP and H323. This translates into faster time to market for new services: New services that took six to twelve months, or more, to incorporate into the old proprietary Class 5 switches can, thanks to new technology, be implemented by third parties physically removed from the switch in a matter of days or weeks.

Chapter 2 includes a descriptive list of competing technologies for the Wireless Internet. We discuss the major technologies presently in use as well as some enabling technologies that will contribute to the success or failure of this new market. A description of the devices available and how they may be used, today and tomorrow, is presented. This chapter is not meant to be a definitive reference document for engineers or professional developers; rather it is meant to highlight the distinctions between those technologies that are the driving force behind the Wireless Internet.

Many applications for data communications have been created such as CDPD, Mobitex, Ardis, RIM, and Ricochet just to name a few. Some of these applications work on standard analog cellular phones whereas others work on custom wireless devices. With the advent of newer digital devices, many new competing protocols have been created for data transfer such as SMS, WAP, i-Mode, c-HTML, and J2ME. The relationships among these and XML are discussed. We also discuss some wireless technologies such as Bluetooth, 802.11, and Home RF and how they fit into the Wireless Internet.

The wireless network infrastructure will be based on many new protocols such as SIP, H323, Megaco+, and others. A description of some of these protocols and their relationship with the new technology is presented as an introduction to the next generation of wireless networks.

Sizing the Wireless Internet market is not an easy task, with estimates from various consultants spanning a vast range. To better understand these projections, in Chapter 3 we have studied research from leading consultants and have taken into account information from middleware players, carriers, and handset and device manufacturers. We believe that the ultimate trajectory of maturity that the Wireless Internet will follow is directly linked to the success of the wired, HTML-based Internet. As such we chart out the growth path: Simultaneous with the growth rate in overall Internet traffic will be the increase in market share of that traffic running over wireless Internet devices like cell phones, pagers, and PDAs. The consensus view is that in the next few years, wireless devices, not the desktop computer will make up the majority of connectivity to the Internet. We explore how the United States, European, and Japanese mobile phone markets are becoming increasingly seeded with data-enabled handsets and devices. We discuss the growth of subscribers as well as exploring the future potential of machines, automobiles, and appliances that will come embedded with wireless communications links for data exchange.

We discuss market forecasts in overall Internet traffic, drivers to wireless penetration in developed and developing countries, rates for wireless services offerings, market growth for data enabled handsets and devices, and some of the pertinent human factors from behavioral, usage, and acceptance perspectives. We provide realistic and credible market information and business implications about what operators and users can expect in the near future. We cite leading market research reports and leading industry experts, with the intention of bringing some clarity to business and consumer market issues in the Wireless Internet.

Applications will be the critical driver for the success of the Wireless Internet, with companies, service providers, device manufacturers, and content developers focusing on identifying and developing the "Killer App." Chapter 4 provides an overview of wireless services and key applications. With the convergence of communications and computing evolving into the next generation, the Wireless Internet is transforming our lives—how we work, how we entertain ourselves, how we conduct business, how we live. We identify and list these opportunities regardless of the underlying technologies—opportunities that provide new information content services and applications tailored to a mobile lifestyle, relevant to personal needs and preferences and accessible anytime, anywhere.

As the Wireless Internet adds new value to staying connected, short response times assure the validity of information. Productivity is no longer confined to a specific location. We explore how the Wireless Internet's new and expanded use of radio waves results in new opportunities, and revenue streams for carriers, and provides the handset and device manufacturers with new markets. We believe that the launching of Wireless Internet service will be a major boost for the global information and communications industry, resulting in hundreds of new venture capital–funded companies and thousands of new jobs. In this chapter we provide many examples of these applications, such as multimedia messaging, which makes it possible to combine conventional text messages with richer content types—photographs, images, voice clips, and video clips.

Another example depicts two of the fastest-growing industries in the world—entertainment and mobile communications—that will profit hugely as lifestyles change and people have more free time. Fast access to entertainment is increasingly appealing to all sectors of society. Many wireless handsets and devices are already used for entertainment. Just as SMS services led the revolution in entertainment on the move, we are now on the edge of a new era as the Wireless Internet begins to offer more sophisticated services.

We also focus on personalized and location-based services, content, and applications that enable users to have richer, more rewarding, and more relevant experiences. The importance of the wireless device as an instrument for information, entertainment, and transactions will increase as physical boundaries dissolve. We identify many of these new services and the wireless devices needed to access them.

A large part of the success of the Internet has come from the ability to connect with users on a wide range of devices. Chapter 5 explores interoperability issues and some of the current problems surrounding the various technologies. Wireless Internet standards seem to be anything but standard today: Each manufacturer views implementation differently, and this leads to consumer confusion when no two devices work the same ways. (In fact, many models manufactured by the same company have different characteristics in user interface and feature implementations.) This chapter deals with the new challenges that mobility adds to the ability to exchange data between devices and systems that lack common input and output capabilities.

Although new advances in wireless computing and applications excite us, we all know that none will survive without a secure revenue stream and an acceptable billing strategy. Privacy and security will be crucial to the uptake of value-added access and m-commerce services. This chapter discusses the need for digital rights management and the kind of distribution that will be enabled by secure and protected content.

With technology also comes threats to the social fabric of our lives. The idea of Big Brother watching over us has never been more accurate. The advent of Global Positioning Service (GPS) combined with two-way mobile communications has created new marketing opportunities for location-dependent advertising. Many companies are "mining" the data resulting from GPS location information and other Wireless Internet activities to create an advertising profile on the mobile user. The social and legal ramifications of these issues are discussed in this chapter.

The future is about communications but communications has always been about the future! The future is not only about great technologies but also equally about how and why consumers and businesspeople use it. Chapter 6 explores the social and emotional impact possible not from technology itself, but the real world use of it.

Consumers are used to the fast pace of wired Internet development, where functionality and expectations are constantly changing. The capabilities of the wired Internet influence expectations of the Wireless Internet. Technologies such as SIP, IM, and DSL are changing the way consumers use the wired Internet and will influence their expectations for the Wireless Internet and those companies that provide it.

The Internet is a powerful economic force, and access will play a large role in determining the growth of developing countries seeking to overcome the Digital Divide. This chapter discusses some of the ways the Wireless Internet can do its part in turning the Divide into a Digital Dividend.

As networks and devices change so will consumer expectations. The primary trouble is that consumer expectations often run ahead of network and device ability. We explore the advent of user-aware systems that adapt content and services to the current context of the user. Wearable computing is ahead of most consumer expectations, and so we give a glimpse of what the future might include.

The future of the Wireless Internet is certain—certain to change in ability and in expectation. We discuss how these two perspectives are likely to cross paths and produce a version of the Wireless Internet that both consumers and industry leaders can embrace, if even for a brief moment before definitions and expectations shift again.

And finally, will the Wireless Internet survive? Our opinion is a resounding "Yes," but like other technological innovations, it is certain to change with time and experience.

A glossary is included for quick reference of terms and TLA's (three letter acronyms) found in this book. For clarity, a brief explanation of each term is included.

Included as appendices are:

- A—Reference library for further reading
- B—List of interesting Wireless Internet and industry information URLs

All Web sites listed in the Appendices will be included on a links page in a special section of our site at www.genesiscampus. com/appendixb.htm. This will provide the reader access to more information than the authors could ever hope to include in a book. These links will be updated with new information regularly, creating a "living" reference document. The host Internet sites will regularly be updated with new information as the technology evolves but the authors of this book can neither guarantee that these links will always be available or as to their accuracy of content.

The authors hope that each reader finds something useful in *Wireless Internet Crash Course*. The Wireless Internet is indeed a "hot topic" in the wireless communications industry. The greatest single technological feat of the 90s was the creation of the Internet. People from around the world, people of every profession, people rich and poor, now have nearly unlimited access to every piece of data that has ever been collected and written down, and they all have equal access.

Today, we are a wireless-enabled society, deeply saturated with mobile devices. Traditional methods of communications and access to data cannot be applied. No longer is the individual tied to an office, home phone, or computer. This untethering of the Internet abounds with new solutions and opportunities for human communications offering wireless connectivity to the vast knowledge and resources of the Internet.

ACKNOWLEDGMENTS

The authors could not have completed this book without the help of many people. The authors wish to acknowledge those whose help and advise were paramount to our completion of the task. We appreciate all of the confidence, encouragement, support, and patience of our editor, Steve Chapman of McGraw-Hill Professional Book Group and to Patty Wallenburg for putting up with all of our editing changes. We also thank Lawrence Harte at APDG Inc. for encouraging us to embark on book authorship, Sudhir Gupta and Jane Bixler at Spatial Wireless for their keen insight into the world of IP, VoIP, and softswitch technologies, and Ari Reubin and Jeffrey Wolf at Sensatex for the look into personalized, mobile information processing.

We wish also to thank Alastair Westgarth of Navini Networks, B.J. Rone of Tatum CFO Partners, Harry Blount at Lehman Brothers, Allan Coon at Anritsu, Monica Paolini at Analysys Research, Steve Sievert at Compaq, Cherie Gary at Nokia Mobile Phones, Wendy Roberts at Kyocera Wireless, Robert Elston at Ericsson Inc., and Juli Burba at Motorola.

We would also like to extend a special thanks to the city of Richardson, mayor Gary Slagel, and city manager Bill Keffler.

Also our gratitude goes to the Technology Business Counsel, particularly Ron Robinson, John Jacobs, and Mike Chism for their comments and encouragement. Finally, we could not have completed this book without to support of Wu-Fu Chen and Junli Wu at Genesis Campus.

HISTORY OF
MODERN
COMMUNICATIONS

It could be argued that the Information Age began in 1837 with the invention of the telegraph in the United States. The first public telegraph was completed in 1844 and ran 64 km or about 40 miles between Washington, D.C., and Baltimore, Maryland. Obviously, Samuel B. Morse was aware of his place in history when he transmitted the first message, "What hath God wrought?"

Morse realized in that instant that communications between individuals and nations had been dramatically altered. Today we take the first steps toward another milestone—the Wireless Internet. To understand the significance of a Wireless Internet, we should look at some of the milestones along the way. It has been said that "Rome wasn't built in a day," and the Wireless Internet will not happen instantly either. It has taken 164 years to get this far—from the invention of the telegraph to today's Wireless Internet.

In this chapter we review a little of the history of wired and wireless communications and the reason for its progress. Technology can drive applications but sometimes, applications create a need for new technology; thus it is with the Wireless Internet—one must understand that having the capability does not mean that the capability will be used. This chapter introduces some of the technological achevements that will culminate in the Wireless Internet:

- Voice and data communications
- Birth of the cellular telephone
- Wireless communication devices
- 2G and 3G cellular
- Technologies driving the Wireless Internet

FIRST VOICE COMMUNICATIONS

Voice communication became possible when Alexander Graham Bell invented the telephone on March 10, 1876. His experiments with his assistant Thomas Watson finally proved successful when the first vocal sentence was transmitted: "Watson, come here; I want you." The telephone was demonstrated to the world at the 1876 Centernial Exposition in Philadelphia, Pennsylvania, and led to the creation of the Bell Telephone Company in 1877.

By 1906 American inventor, Lee De Forest, invented a three-element vacuum tube that revolutionized the entire field of electronics by allowing amplification of signals, both telegraphy and voice. The first radio broadcast in the United States was made in 1906, and within four years the first broadcast from the Metropolitan Opera House was transmitted.

Wireless voice communications using amplitude modulation (AM) was a reality. The ensuing years of the 1920s saw tremendous growth in radio station broadcasting that brought the possibility of real-time information to the public. Society changed forever...again. The radio became a necessity for people to communicate information and ideas over vast distances without wires.

Of course, wires still had their place because radio was not always the most reliable medium. The environment, weather, time of day, and man-made interference could interrupt communications. Telephone technology advanced steadily, and telegraphy still found a place in data communications in the form of the *telegram*.

Radio technology advanced throughout the 1930s with the notable invention of frequency modulation (FM), which provided better sound quality and was more resistant to interference than the older AM broadcasting system. One of the first applications for FM was police radio; it was ideal for mobile communications. Commercial FM broadcasting did not develop until much later in the twentieth century. It should be noted here that FM technology became the cornerstone of the analog cellular system launched in 1983.

World War II accelerated the advancement of radio communications and electronics. Transatlantic cables between Europe and North America improved but we were still limited to real-time communications by copper cables or high-frequency (HF) radio spectrum under 30 megahertz. Data was still limited to telegraphy or some analog signals representing data. This was acceptable because demand for data was also low.

However, the post-war period saw an explosion of innovation with the development of the transistor (December 1947) and the birth of the computer. In the Moore School of Engineering, ENIAC, the world's first electronic, large-scale, general-purpose computer, was activated at the University of Pennsylvania in 1946. Unfortunately, the computer preceded the transistor so ENIAC contained about 18,000 tubes. This was much to the chagrin of the graduate students who had to replace the burned out ones—often! Some refer to this as the Birth of the Information Age, but we like to think of it as the Re-Birth of the Information Society. Computers provided a tool for people to process data, lots of data; now we needed a better way to move that data faster.

The 1950s had many "Ages" to ponder, the Atomic Age, the Information Age, and if that was not enough, another almost 100 years after the first transoceanic cable, another society-altering event occurred, one that changed the way we communicate and, perhaps even more so, the way we think globally. The Space Age began with the launch of the Soviet satellite Sputnik on October 4, 1957. Satellite communications provided reliable long distance communications by augmenting or replacing cables. This created the demand for reliable, anytime, anywhere communications. The beginning of an idea for a truly mobile, global society was planted; the capability to link people around the world with nearly instantaneous voice and data communications was a reality, but it was still fixed point-to-point communications.

The Space Age brought changes to the way we think and the technology we create. It brought us integrated circuits, fiber optics, photonics, ceramics, freeze-dried food, and ultimately digital electronics. Digital technology enabled the creation of computers, as we know them today, and the transmission of data at higher speeds. It also provided wireless, high-bandwidth communications. Communications satellites and transoceanic cables—including technologically advanced fiber optic cables with high bandwidth—continue to be installed around the world.

It took almost 26 years after Sputnik before cellular communications brought mobile voice communications to the masses (at least those who could afford $4,500 for a mobile phone in December of 1983). Mobile data took a couple more years to become common, but speed and reliability remained issues to its success. Outside the military, access to large databases of information was still limited to commercial and educational institutions with their internal mainframe computers. (Because sharing this data over wireless connections has been impractical, data networks have remained mostly wired.) During the 1980s, no compelling need for wireless data transmission existed. That was about to change…. Figure 1-1 illustrates the communication timeline.

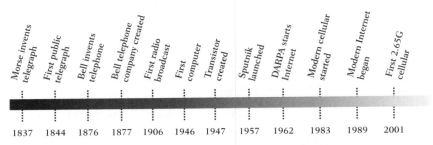

FIGURE 1-1   Communications milestones.

CELLULAR, TRUE MOBILITY FOR THE MASSES

Meanwhile in a separate segment of communications, another revolution was taking place—cellular telephones. In the late 1970s AT&T Bell Laboratories began working with several leading United States and Japanese companies to create a cellular telephone system based on dividing coverage areas into small cells and reusing frequencies. Previous mobile telephone technologies operated on limited numbers of channels, thus limiting the number of users in any given coverage area to a very small number. The result was low user use and costly service and equipment. A core group was created to develop a standard called the Advanced Mobile Phone Service (AMPS). In December 1983, AMPS was launched in Chicago, Illinois with great fanfare. It proved immensely popular. Now before someone says, "Hey, wait a minute, AMPS wasn't the first cellular system!," let's give that credit to the Nordic Mobile Telephone (NMT) system. NMT was launched in 1981 in Scandinavia, but in terms of market size, AMPS potential market in the United States was vastly larger. AMPS quickly spread to other countries in North and South America, Korea, and Australia. A similar standard, Total Access Communications System (TACS), was developed in the United Kingdom as well.

Today, there are many competing standards in mobile telephones worldwide. In fact the word "mobile" means something entirely different today than it did in 1983. The majority of cellular telephones sold today are hand-held, not permanently installed in vehicles. Each competing standard is incompatible with others on the basic technology used, but to the end user, all cellular telephones should perform the basic functions expected. (Even though many new carriers would like to distinguish themselves from "cellular" companies by calling themselves "PCS" companies, we consider both as cellular applications in this book. This is not to say that companies with PCS spectrum in the 1900 MHz band may or may not have some advantages over carriers with traditional spectrum allocations in the 800 MHz band. But because many carriers own spectrum in both bands this is a moot point.)

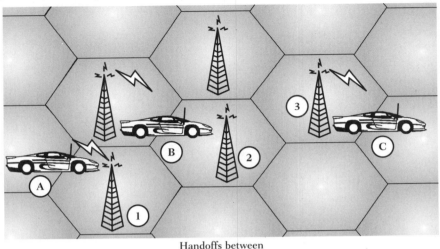

Handoffs between
cells sites during
travel from A to C

FIGURE 1-2   Cellular concepts.

Cellular radio got its name from the physical layout of a system in a pattern resembling a honeycomb figuratively. In Figure 1-2, a vehicle traveling from point A to C, will initially be communicating through cellsite 1. As it moves to position B, communications is handed off to cell site 2 and similarly for position C. Each cell site will operate on a different frequency so that neighboring cells do not interfere with one another. However, frequencies can be re-used if they are separate by sufficient distance. This is referred to as the re-use pattern.

COMPUTING POWER IN MOBILE COMMUNICATIONS

Computing power can mean many things depending on where the term is applied. The computing power of today's cellular handset is much greater than just ten years ago. The mobile phones of the 1980s used 8-bit microcontrollers with very little memory. A typical phone operated with 6 Kb of RAM (scratchpad memory) and 32 Kb of ROM (program memory).

That was fine, because AMPS was an analog communications system based on FM and the data rate was low because all data requires a modem to convert digital data to analog modulation. Data rates for wire-line modems in the early 1980s were initially 300 baud progressing to, at best, 19,200 baud by 1990. Therefore, rates of 1200 baud to 2400 baud were acceptable for wireless device communications. (One important point should be noted—gross baud rates and throughput are two different things entirely!)

The wireless cellular communications channel is a dirty, nasty place for data communications signals. Impairments to an analog cellular channel are noise, weak signals, interference, and signal dropouts caused by handoffs from one channel to another. Today's digital cellular channel problems are compounded by signal degradation, multipath fading, and delay.

But today's cellular phones have more computing power than the average workstation of the early 1990s. They will contain 16-bit, 32-bit, or 64-bit RISC microcontrollers, digital signal processors capable of 400 MIPS or more by 2002, and they will contain megabytes of memory. This translates to computational power sufficient to make a digital voice call or send high-speed data such as full motion video while simultaneously reading your email on the display. Data rates will soon approach ISDN or DSL rates and may go higher, anywhere from 114 Kbps to 2 Mbps or higher.

This vastly increased computing power (and the interest of the Defense Department) has brought one other very important new feature to wireless communications devices—Global Positioning Services (GPS). Using GPS, a wireless device can communicate its location to anywhere in the world. Orwell look out: Location-based marketing to a mobile customer base is coming.

THE INTERNET, A NEW IDEA

The 1990s saw the emergence of the Internet as a dominant communications media but actually its beginnings can be traced back to the late 1960s. The Internet started out as an idea born within the Rand Corporation, America's premier think-tank for Cold War strategy. The Defense Department wanted to create a method for communications of defense command and control information in a post-nuclear world. It required a decentralized network that could function even if several nodes were destroyed. Ultimately, a request for proposal was issued from the Defense Advanced Research Projects Agency for a packet-switching network that DARPA was planning to build. Several graduate students and faculty at the University of California–Los Angeles presented a proposal to DARPA for establishing a communications network model known as the ARPANET project. There were no formal standards written for ARPANET, so a method of documentation was devised: the RFC or *Request for Comment*. (This method continues today and throughout this book, references to RFC numbers will be made.) By December of 1969, a four-node network was working and by 1972, thirty-seven nodes were working in the ARPANET. The network enabled researchers across the country to share computational resources.

Somewhere along the way, an interesting observation took place that the ARPANET was really a government subsidized person-to-person communications service more than a sharing of resources. The advent of personal user accounts enabled an electronic mail service. Shortly after this the *mailing list* was invented, which enabled the broadcasting of messages to large numbers of users simultaneously. Researchers could now communicate personally or share ideas with a group. ARPANET in its infancy was shared by academic institutions and their financial backers, the Department of Defense. The network grew throughout the 1970s because of its ability to add nodes using many different computers as long as they "spoke" the same language of ARPANET. The original language was *Network Communications Protocol* (NCP) but it was superseded by TCP/IP in the early 1980s. *Transmission Control Protocol* (TCP) breaks each message into packets at the source and then reassembles them at the destination. Each packet contains a source and destination address so that *Internet Protocol* (IP)

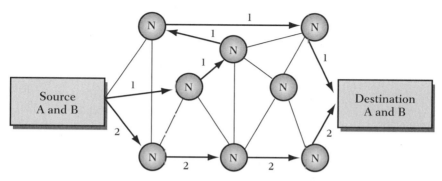

FIGURE 1-3   Packet data network principles.

can route the packets through multiple nodes and multiple networks successfully, even if the nodes or networks operate with different standards.

The principles behind TCP/IP and the Internet are that data can go many different paths. Figure 1-3 shows how two different data sources, A and B, can travel along separate paths 1 and 2, through multiple nodes (n), and them come together at the destination.

Concurrent to the development of this vast network of supercomputers was the invention of the personal computer in the early 1980s. Suddenly thousands of individuals had access to a computer on their desktop or in their homes. This played heavily in expanding the ARPANET although it remained closely controlled until 1983, when the Defense Department split off the MILNET. By this time many groups of people had access to computers and through the simplicity of TCP/IP public-domain protocols, they could link to the network and essentially add another node. Thus the "network of networks" was created which ultimately became known as the *Internet*.

By 1984 the National Science Foundation (NSF) jumped into the fray, promoting technological advances for ARPANET. Faster speeds were achieved through upgraded links and newer supercomputers. Other government agencies joined in expanding the network and the cumulative knowledge base of information. A method of identifying users was devised to create *domains* with unique identifiers such as com, gov, org, mil, edu, and net. At the same time two-letter country designations such as uk, dl, and fr were created because the network now crossed international boundaries. By 1989, ARPANET passed into history, a victim of its own success. The birth of the Internet had occurred as a result of many people's long, arduous hours of research and development. It was not invented by a politician as some would like us to believe!

The 1990s saw explosive growth in the Internet, with hundreds of companies supplying enabling technology, thousands of companies providing service, and even more users sharing the combined wealth of knowledge on the Internet. By 1993 there were over 1.3 million computers connected to the Internet. Today there over 20 million hosts and 500 million users, sharing ideas and knowledge, swapping emails, and buying and selling through e-commerce.

The explosive use of the Internet during the 1990s could be compared to other momentous events such as the invention of the wheel, the Industrial Revolution, and the invention of the transistor or integrated circuit. No single "invention" or "revolution" has affected our lives more than the Internet. A whole new generation has grown up with access to the 'Net, accepting its promise to communicate with anyone, anywhere. That same generation considers the cellular phone a necessity and would not go anywhere without it.

Although the original intended use of the Internet was file sharing and electronic mail, it soon became apparent that it really was a tool to connect people to people. It created the world's largest, easily accessible marketplace and gave birth to e-commerce. The under-25 age group represents a huge market segment, and an entire industry has grown-up in the "dot com" market segment catering to the needs of these Internet users. However, until recently, this was a "tethered" connection. If the Internet could be extended beyond wired connections, it could be accessed anytime and anywhere.

This brings us back full circle to wireless technologies and raises the question, "How can I access the Internet from where I am at this moment? How do we put all of this together to benefit us?" Let's consult our crystal ball to look into the future. We know that the Internet will expand our horizons and opportunities for new services. We also know that wireless communications is becoming very popular and that data rates are increasing. Added to all of this, we can know wireless user location thanks to GPS. An application environment to connect the wireless user to the Internet is ready. All that the Wireless Internet needs are applications to drive user adoption. For service providers, developing and launching these applications is a gamble. If user satisfaction is low, the Wireless Internet will be a huge disaster. The financial implications will be staggering to the world economy, considering that service providers have committed billions of dollars on licenses, infrastructure, and development all predicated on the success of the Wireless Internet.

WHAT ABOUT THE WIRELESS INTERNET?

Early attempts at wireless data transmission evolved around proprietary technologies. Usually, the cost to deploy such networks limited their use to that of large companies. The services offered usually revolved around some form of dispatch service. Today, however, cellular networks are ubiquitous and quite capable of serving the data requirements of not only large companies, but individuals. Cellular is changing the way we communicate on an everyday basis.

Another factor in moving to a Wireless Internet is the size of computing devices. Miniaturization and improved batteries are providing smaller, better mobile tools. Laptops and PDAs are small enough to be very mobile but powerful enough to tackle anything that we might do on a desktop computer. Portability and connectivity can be readily achieved. Now a mobile businessman can be more productive because he can access his data in his office or retrieve data stored from a global Internet connection. As the performance of the Wireless Internet approaches that of a fixed connection, there is no longer a need to remain tethered to a desk.

There are two types of Wireless Internet connections: those
through cellular and those through a mobile data network.
There is a very big distinction between these two types of connection. Cellular has traditionally been a circuit-switched connection, whereas mobile data networks are packet based. The
next generation cellular standards will eliminate this difference. Data on the Wireless Internet will be packet-based using
TCP/IP, the same protocol used on the Internet. The spectrum
resource or channel will become a shared resource, and new
methods of *usage billing* rather than *airtime billing* will emerge.
Data speed will also increase from 64 Kbps to more than 2
Mbps depending on the technology.

The Wireless Internet is a natural and inevitable progres-
sion from the wired Internet when you consider today's wireless communications devices, a very mobile society, and a free
market economy where anything can be sold if it has the right
sales approach. Cellular penetration is very high, with over 1
billion cellular users projected by the end of 2002. Some estimates put data revenue streams in 2006 higher than today's
voice revenue streams. This may be a bit optimistic but it is
clear the demand for wireless data transmission is growing. A
cellular device is a personal device and the value of wireless
data is in the knowledge of the user, his buying habits, his location and other personal information. The proper use this
knowledge will create new revenue streams. Applications must
be created that the user cannot live without.

The terminal market varies to users' demand. As we stated earlier, a Wireless Internet device does not have to resemble a cellular phone or even possess the functionality of a cellular phone.
The new Wireless Internet will be accessed by many new devices
and methods. Voice functionality does not necessarily need to
reside on the device. Voice recognition and text-to-speech may be
the solution for access. User interfaces must change to reflect the
wider spectrum of data throughput. Displays and keyboards may
no longer be of primary importance: If you have voice recognition
and text-to-speech capabilities, do you need a bigger display and
keyboard? Maybe the Wireless Internet device will require no
human interface. Technologies used in terminal devices are large-

ly determined by the application required. (In Chapters 4 and 5 we discuss this subject further.)

WIRELESS COMMUNICATIONS DEVICES

Today's wireless communications devices are not your father's car phone of yesterday. Mobile telephones in the 1980s were big and heavy. A phone's weight was measured in pounds instead of ounces. Those that were hand transportable were referred to as "luggable." They were limited to low resolution character display.

Cellular phones today are portable, hand-held devices smaller in overall size than yesterday's mobile phones. Personal Digital Assistant (PDAs) are more closely related to a small computer than a cellular phone. Some include a proprietary radio frequency modem for wireless communications whereas others simply connect to a cellular phone through a serial cable, radio link, or infrared port.

Figure 1-4 is a picture of an older "portable," large by today's standards. Yet the first portable or "luggable" device were many times larger and heavier than today's handhelds. One of the earliest units was consisted of a full three-watt mobile like the ones installed in car trunks, a phone handset almost the size of a home phone and a twelve-pound nickel-cadmium battery. The whole unit was mounted in a heavy-duty transportation briefcase and tipped the scales at forty-five pounds!

Figure 1-5 shows several manufacturers' wireless PDAs and Pocket PCs while Figure 1-6 illustrates some manufacturers' current models of handheld cellular phones.

Tomorrow will bring many new wireless devices that communicate over commercial cellular or within their own assigned frequency bands. An example might be a *smart* or *Internet appliance* that communicates warranty information to the manufacturer or requests service. This will all be done over the Internet, with a message returned to the end user to notify him that the appliance has reported a malfunction and will require service or that a software upgrade was completed and no further action needs to be taken.

FIGURE 1-4   Yesterday's cellular phone. Photo courtesy of Oki Telecom.

No matter what the end application, two things are rather clear:

1. Any information presented across the Internet to or from wireless devices will be limited by network bandwidth and the display capabilities of the terminal device. To the end user, bandwidth is the quantity of information transmitted per second. In the wireless world, this is limited by the size and efficient use of spectrum. A wireless device will not display all of the information that a traditional Web site may contain. Either some type of filtering will be applied to convert Web site content to a suitable size, or other user interfaces will be developed—such as text-to-voice translation or a dedicated wireless Web site.

2. The projected number of personal computers connected to the Internet will be exceeded by the number of Web-

FIGURE 1-5A   Wireless PDAs. Courtesy of Palm, Inc.

FIGURE 1-5B   Wireless Pocket PCs. Courtesy of Compaq Computer Corp.

FIGURE 1-6A   Cellular handsets. Photos courtesy of Nokia Mobile Phone.

FIGURE 1-6B   Cellular handsets. Photos included courtesy of Ericsson, Inc. ©
Ericsson 2001.

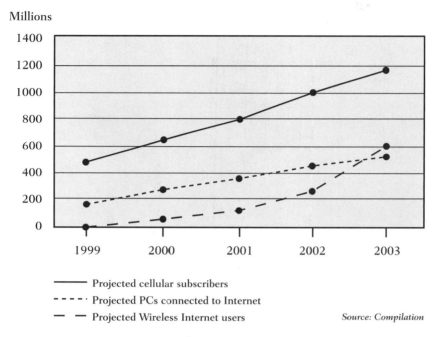

Millions

Projected cellular subscribers

Projected PCs connected to Internet

Projected Wireless Internet users

Source: Compilation

FIGURE 1-7   Internet connectivity outlook.

enabled handsets by about 2003, see Figure 1-7 (600 million wireless Internet users: source Dataquest).

The only way to currently get Internet content into most of
today's Wireless Internet devices is to "squeeze" content as
shown in Figure 1-8.

TRENDS IN NEXT-GENERATION MOBILE COMMUNICATIONS

Shortly after cellular first launched, third-party entrepreneurs
began offering modems for cellular data transmission. These
modems resemble the standard analog modems circa 1985.
They transmitted data at 300 baud, used a special cellular protocol called MNP-10, and an *AT command set*. MNP-10 made
these modems different from regular modems because they
had to be fault-tolerant to cellular hand-offs and the in-band
signaling tones used in analog cellular. One such modem was

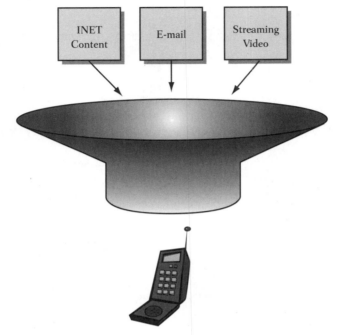

FIGURE 1-8   Squeezing content.

created by a Dallas, Texas company, Spectrum Cellular, that actually consisted of a modem set or pair. The land or fixed side modem was installed at the cellular switch. The mobile half was connected to the mobile phone and computer. The connection was circuit-switched, which meant that the phone was connected as long as data was transmitted and the cellular traffic channel was dedicated to that one user.

The next milestone for data over cellular was Cellular Digital Packet Data (CDPD) in 1992. An industry consortium of leading wireless communications companies set out to develop the CDPD specification. Their design objectives were to send digital data over the existing AMPS wireless infrastructure without major changes to the AMPS infrastructure, with reasonable performance (19.2 Kbps), high reliability, and security; and to support seamless roaming. CDPD also had to co-exist with voice traffic. Because it was based on standard Internet TCP/IP protocols, service providers that do not have next-generation digital solutions for consumers today can still use CDPD as an Internet connection. The CDPD market is alive for the time being, but this may change as next-generation networks become active.

THE MOVE TO 2G CELLULAR

In 1989, the cellular industry began the task of migrating cellular from an analog technology to one of several digital technologies, primarily to increase capacity in several cities that were in danger of running out of voice capacity (as in New York and Los Angeles). Data was not the overriding concern for the standard bodies, and unfortunately, not every region adopted the same migration path. This created a challenge to the network designers working to maintain uniformity of operation.

The European community chose collectively to migrate their existing networks to Global System for Mobile Communications (GSM) for their first-generation digital networks. Meanwhile North America chose to develop a digital standard in two parts. The first was referred to as Interim Standard-54 (IS-54) or North American Digital Cellular. This standard, based on a version of Time Division Multiple Access (TDMA), is similar to GSM but incompatible. IS-54 was developed during the early 1990s and was soon followed with IS-136. The difference between the two was that IS-54 continued to use analog control channels and used both analog and digital traffic channels. IS-136 contained both digital and analog channels for control and traffic. IS-54 was not widely popular because it lacked clear advantages to the user. The promise of greater battery life with IS-136 alone was inducement enough to win customers over even had it lacked other advantages.

Having two different network architectures wasn't too bad but wait—there's a new show in town. Here comes a company out of San Diego that no one has ever heard of before, and they claim to have a better solution to digital cellular. The company was Qualcomm and the solution proposed was Code Division Multiple Access (CDMA). Today, Europe still has GSM and North America has TDMA, CDMA, and a little

GSM for variety. By 2001, GSM occupied about 65 percent of the worldwide cellular market and CDMA held about 17 percent market share.

As far as wireless data is concerned, we have only three choices: CDPD on an analog channel, Short Message Service (SMS) if it is available, or an internal modem in the phone or device. Each choice has its pros and cons. Short Message Service is just that—short—so speed is very important to the user. The rate for CDPD is 19.2 Kbps, which is fine for certain text applications, and because it is packet-based, valuable spectrum is not wasted. The third option is a built-in modem in the phone that connects to a laptop computer by either cable or infrared. Speeds with an internal modem range from 8 Kbps to 9.6 Kbps. The data simply replaces the voice traffic transmitted by the phone; the connection is circuit-switched, so spectrum is wasted. None of these options is satisfactory for real-time access to the Internet or streaming video.

ONWARD TO 2.5G AND 3G

The next step to higher data rates for each technology was dubbed 2.5G and was to be closely followed by 3G. (The "G" of course stands for "generation.") 3G is not just a standard for higher data rates: It is also meant to bring global standardization to cellular. Our choice of words here is very deliberate: "closely followed" has been defined by some as within two years of 2.5G, whereas others say that the two standards are practically on top of one another. The simple fact is that it costs a lot of time and money to upgrade a cellular system and it may make more business sense to skip interim steps. Just as IS-54 was quickly replaced by IS-136, carriers may find 2.5G unpalatable financially. In other words, they may skip 2.5G and go directly to 3G. That makes sense but what happens when industry skips an interim solution or worse yet, adds another? Do we add a 2.75G?

The truth is that in the interests of harmonizing all of the different proposals for 3G, the cellular industry has skipped some steps or in some cases, changed direction altogether. Like

Qualcomm, NTT DoCoMo, the Japanese telecommunications giant, has proposed a new standard altogether, Wideband Code Division Multiple Access (WCDMA). Three years ago, the roadmaps for GSM, CDMA, and TDMA were clear. GSM would become GSM Phase 2+ with improvements and the addition of High-speed Circuit Switch Data (HSCSD) and rates up to 144 Kbps. Later, it would migrate to Enhanced Data Rate for Global Evolution (EDGE). GPRS would be added along with a more robust modulation scheme; rates of 384 Kbps would offer wireless multimedia IP-based services and applications. At that time there would be an "alignment" with TDMA; each would have an EDGE physical layer. Figure 1-9 shows the most recent roadmap for 3G.

Meanwhile, TDMA would become IS-136 Plus with the addition of HSCSD. It would then migrate to IS-136HS (EDGE) just like its cousin, GSM. (Not all EDGE is created equal: The European version of EDGE and the North American version share a common standard but different frequencies. A "world" phone would have to cover more bands in order to roam.) The roadmap for GSM and TDMA primarily increases data capabilities, not voice. It is expected that voice transmission would migrate to Voice-over-IP in the future. Until that happens, EDGE is split into two component networks, one for voice and one for data.

The CDMA side was also quite clear three years ago. IS-95A would become IS-95B with HSCSD up to 64 Kbps. Later IS-95C and IS-95D (sometimes referred to as IS-2000 Phase I and II) would increase data rates to about 307 Kbps. IS-95C was also referred to as 1xRTT, and IS-95D referred to as 3xRTT. Just to confuse things a little, another standard was created to overlay the other two. High Data Rate (HDR) could go as high as 2.4 Mbps. (The latest acronym for 1xRTT combined with HDR is 1xEV.)

Well, today most of this has changed. NTT DoCoMo proposed a new Wideband CDMA. When the technical and political ramifications were viewed, deals were made between proponents of each standard. GSM Phase II+ survived but IS-136 Plus didn't. For CDMA, 1xRTT survived but IS-95B didn't

22

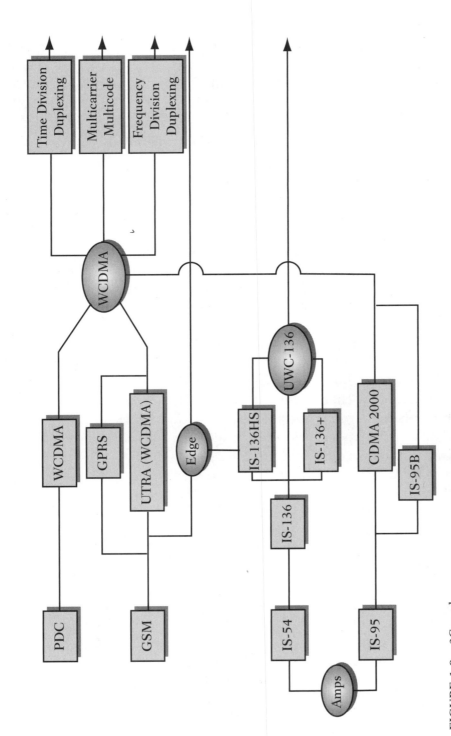

FIGURE 1-9   3G roadmap.

and 3xRTT will most likely be delayed as long as HDR meets users' needs. As for WCDMA, it will become a component overlaying GSM networks, transforming those networks into Universal Mobile Telephone Service (UMTS).

The last three years have been very confusing to those following the standards process. The only things that remain clear are that CDMA will launch 1xRTT with an HDR overlay to support date rates from 144 Kbps to over 2 Mbps. Both data and voice capacity will benefit from 1xRTT.

Europe will launch EDGE with WCDMA overlaid. Data rates will run from 384 Kbps to over 2.5 Mbps. Both data and voice capacity will be improved. In the United States, the IS-136 component of EDGE may be eliminated in favor of the European version of EDGE and WCDMA. AT&T Wireless has announced a decision to do exactly that: overlay the old IS-136 network with a GSM/EDGE/WCDMA version.

In the United States, a fourth cellular technology is deployed by Nextel. It uses a proprietary technology developed by Motorola called iDEN. The Nextel system works as a hybrid design between cellular and dispatch technologies. Calls may be connected like cellular, or members of a group can be connected together in a way similar to two-way radio, without dialing. For roaming outside the United States, Nextel offers a dual mode—iDEN and GSM—phone. For data applications, the Nextel phones include a Java 2 Micro-edition (J2ME) environment and transmit data on a packet network.

Regardless of the details of who implements what, three important things should be remembered:

· High speed packet data will replace circuit-switched data.
· Internet Protocol (IP) will become the standard protocol for all wireless traffic, voice, and data.
· A quasi-global standard will make international roaming easier.

These three changes to mobile communications will open the door to the next generation of wireless applications.

TERMINAL TECHNOLOGIES

Early attempts at data transmission used either an analog modem or CDPD, but these never really proved financially rewarding to the carriers. The analog modems are very slow and do not warrant further discussion. CDPD transmitted packet data over an analog network. It was a niche market rather than a mass market: Data rates were moderate (19.2 Kbps), phones and modems were expensive, applications were very limited, and most people never even knew that it existed. CDPD is still in use as a slow-speed (by today's standards), Wireless Internet connection on analog and dual-mode phones.

Newer equipment and protocols have resulted in many wireless transmission schemes, some competitive (directly or indirectly) and some complementary to the others. For limited mobility applications, we have wireless local and personal area networks (WLANs and PANs) with standards such as Home RF, IEEE 802.11, or Bluetooth. Mobile data networks such as Mobitex and Ardis, for public or private wide area networks are used mostly for dispatch and service industries. Wireless PDAs (Palm, Handspring, etc.) and Pocket PCs (Compaq, HP, etc.) have their own data networks such as OmniSky or they use a cellular phone with a modem. In the cellular networks themselves, SMS, WAP, I-Mode, and J2ME compete as data application platforms. GPRS competes with CDPD or other modem technology on cellular phones. Yet, all of these technologies are needed to make wireless mobility truly ubiquitous. (A more in-depth description of the technical characteristics of these significant technologies follows in Chapter 2.)

With the proliferation of so many standards as shown in Figure 1-10, there is an increasing need for convergence. Users will demand that their Wireless Internet service be simple, fast and uninterrupted. Many locations such as inside buildings are very difficult for wireless carriers to provide adequate coverage.

One possible solution maybe the construction of public WLANs. Wireless LANs are currently being built because the technology is fast, proven, inexpensive and available. Wireless Internet users who operate within a WLAN environment can get better coverage than that promised by 3G. The bandwidth

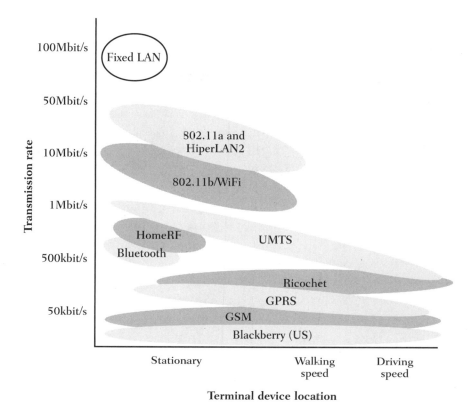

FIGURE 1-10   The Proliferation of Standards. *Source: Analysys 2001.*

available is up to 11 Mbps with 802.11b. Other technologies could result in even higher bandwidths. Solutions will be created that make the experience as simple as possible for both the users and the wireless providers. If billing is handled by the wireless provider on the users existing account, incremental income is realized and the carriers like that rather than viewing the WLAN as a competitor.

INFRASTRUCTURE CHALLENGES

CIRCUIT-SWITCHED VS. ALL-IP

Today's Internet is a packet-based network with an always-on connection. There are very fundamental differences between circuit-switched connections and packet networks. Circuit-switched connections require a real time end-to-end session, whether the transmission is a voice conversation or a data transfer session. Whatever goes into one end always comes out the other in the same order. It may be degraded by static or noise, but its original order is maintained. If there is a loss of voice or data, a repeat transmission can take place instantly to guarantee the reliability of the connection.

Internet Protocol (IP) works entirely differently in the transmission of data, whether using Voice-over-IP (VoIP) or a binary data file. The data is broken up into small entities called *packets* and each one carries a sequence number so that if they arrive out of sequence, they can be reassembled. They can also take different paths to the final destination, so there may be delays in the arrival.

We do not wish to imply that packet is better than circuit switched. The method selected depends on the application. HSCSD allows wireless data to be transmitted at up to 38.4 Kbps or more over GSM networks by allocating multiple time slots to a user. Although this is better than today's average data rate over most Wireless Internet access methods, it will not really support true multimedia content. HSCSD, however, is well suited for time-sensitive, real-time services such as large file transfers. Packet is well suited to short file transfers, messaging, or for longer file transfers where time is not critical.

IP packet-switched networks operate as distributed networks—after all, that was the reason for the creation of the Internet in the beginning. Distributed networks allow for the decentralized control of key elements required of a network such as applications, management, and billing. Packet networks are typically *connectionless* networks. The path that a packet takes through the network can vary from packet to packet. Circuit-switched networks are *connection oriented*. A connection is set at the beginning of a session and remains until the session ends. From a network point of view, connectionless is a far more efficient use of the network resources because resources are shared with all users dynamically. Circuit-switched networks use distance, location, and time as yard-sticks to measure the billing rate for sessions. In packet net-works, distance, time, and location are not as important as the number of packets transferred through the network.

Usage billing becomes far more important in packet net-works. Circuit-switched networks are generally more propri-etary, legacy-based systems, whereas packet networks are much less complex. Service provisioning is far more difficult for cir-cuit-switched networks.

SERVICE PROVISIONING

The whole concept of service provisioning for packet networks and IP billing requires new technology to meet the needs of service providers and customers. Initially, these new technolo-gies augmented those of circuit-switched equipment and later supplanted them. Telecommunications based on Internet Protocol (IP) allow carriers to create grades of service and vari-able pricing to reflect real market conditions.

Currently, data from switches is formatted into billable event detail records through a mediation function. The big dif-ference between the circuit-switched and the IP network is that IP billing must handle many more variables. Just as cir-cuit-switched billing is derived from call detail records (CDRs), IP-based billing is derived from Internet usage records (IURs). However, IURs must contain far greater infor-mation. Future bills will include usage-based billing based on IP information.

Customers can receive value-added information and servic-es with real-time billing. Provisioning of services can take place online for wireless users, satisfying the customer who wants added services. It also benefits the service provider by adding incremental revenue for a given customer and by providing more accurate and timely billing.

To implement IP billing, however, techniques must be developed to retrieve and analyze IP data. Because this holds true whether it's wired or wireless IP, many companies are working on solutions to this problem today.

NETWORK SWITCH REQUIREMENTS

Wireless operators are experiencing a rapid decline in their average revenue per user (ARPU). Strong competition has generated a need for differentiation in operator service offerings.

The advent of the Internet has created a tremendous new and exciting business opportunity for wireless operators. Operators are rushing to upgrade their networks with new packet-based technologies that will allow them to offer innovative wireless data services to their subscribers. This has triggered an immense demand for highly scalable, low-cost, easily maintainable, packet-based, unified, voice and data wireless core switching products.

Wireless subscribers are far more sophisticated users today than they were five years ago. They are no longer satisfied with just placing a call; they require innovative ways to use the wireless phone. New applications for enhanced services are very important to wireless customers. Features such as Caller ID and voice messaging are considered standard. New services and features will become important differentiators in a competitive service-provider market.

To provide these new services and features for the wireless Internet, present-day equipment must give way to new technology. New application protocols must be implemented to work with packet networks.

WIRELESS OPERATOR CHALLENGES.   The core switching network elements found in current wireless networks are called Mobile Switching Centers (MSC). In most cases, MSCs were created by adding wireless-specific interfaces and mobility management functions to existing circuit-switched Class 5 or Class 4 switches. As a result, the incumbent MSC vendors are also traditional switch vendors, such as Ericsson, Nortel, Siemens, Nokia, Lucent, and Alcatel.

Because of the tremendous growth in the number of wireless subscribers and their minutes of use, wireless carriers are continually adding additional capacity to their voice networks. At the same time, to provide data services, new elements (SGSN, GGSN, etc.) are also being added to the network. Furthermore, the evolution of wireless networks to 3G and packet networking—and thereafter to all-IP networks—is leading to the addition of more core switching network elements. The end result is a very complex core network, as depicted in Figure 1-11.

Next generation networks will be comprised of really three types—2, 2.5G, and 3G networks. Each network adds features and therefore requires different interfaces. Compound this with support for circuit switched and packet switched and you have a major headache to connect.

This "patchwork" approach leads to duplicity of functional blocks and unnecessary capital expense (CAPEX) and operation expense (OPEX) costs for the operator. Furthermore, the service logic of voice and data and 2G and 3G remain disparate; this results in slow roll-out of new services. Separate service logic for voice and data also makes it nearly impossible to deliver hybrid multimedia services that require voice–data service integration.

Incumbent core network vendors are not addressing adequately the migration to 3G networks because all their 2G products (as well as some of their 3G products) are still based on old, highly proprietary platforms and their approach to adding new functionality is evolutionary rather than revolutionary.

New products are needed to simplify and streamline the network. A cohesive technology is required to build the next generation, packet-based, unified core switching platform that will satisfy all the voice and data core switching needs of the wireless operator. This new network architecture will lead them to profitability by significantly reducing OPEX and CAPEX and enabling rapid deployment of unique new services.

SOFT SWITCHES AND MEDIA GATEWAYS. Soft switches are poised to replace the call processing functions of the Class 4 and 5 switches previously used by the telecommunications industry. This new breed of switch is smaller, less expensive, less power consuming, more reliable, more flexible, and more efficient than its predecessors. A soft switch can be placed in a closet, whereas the equivalent Class 5 switch functionality would fill a building to achieve the same capacity. The proliferation of control

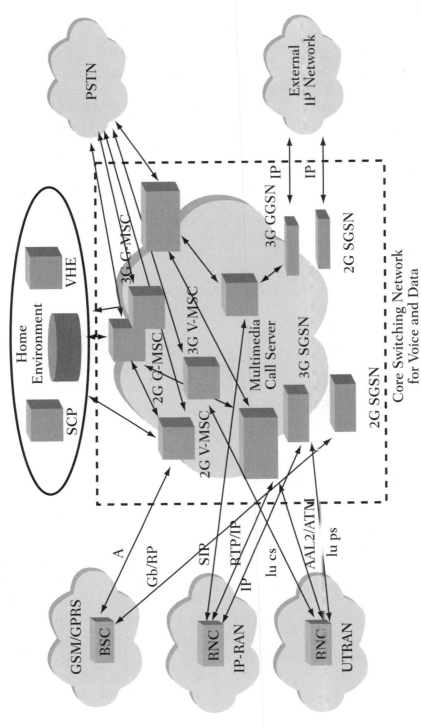

FIGURE 1-11 3G network patches for All-IP.

protocols makes soft switches the ideal solution for today's market. Because soft switches derive their name from the "soft" in software, they can support many protocols at once.

By utilizing soft-switch technology along with a media gateway, a simplified network architecture could be developed (see Figure 1-12). A single core switching product is needed to support both voice and data for 2G, 3G, and future all-IP networks.

The support for all three generations of networks is still a requirement but the interfaces are much simpler.

The solution may be based on state-of-the art, high density, scalable, soft-switch technology and utilize a multimedia session initiation protocol (SIP) session manager as the basic building block for services. This architecture is all-IP ready and fully compatible with 2G, 2.5G, and 3G voice and data specifications. It could support present operator needs yet allow for seamless evolution to future technologies.

A soft-switch/media gateway product developed by Spatial Wireless in shown in Figure 1-13. This product is designed for next generation markets—packet-based core switching for the GSM, CDMA, UMTS and All-IP wireless markets. These elements enable unique voice/data converged services, help maintain service transparency across different wireless generations, and can result in more than 50 percent savings in capital and operational expenditures. The Spatial's Portico product is an overlay gateway product that supports the introduction of voice, data and converged services.

As we have seen in this overview of the history of modern communications, competing standards and protocols both drive and hinder the development of a truly ubiquitous Wireless Internet. Chapter 2 explores the evolution of these technologies in greater depth.

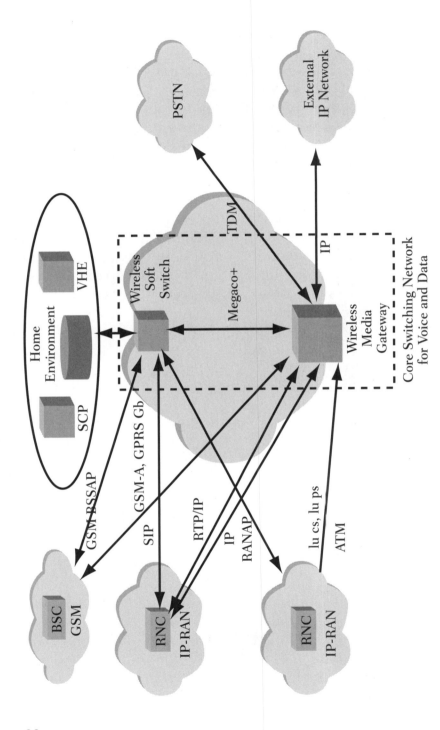

FIGURE 1-12 A wireless solution.

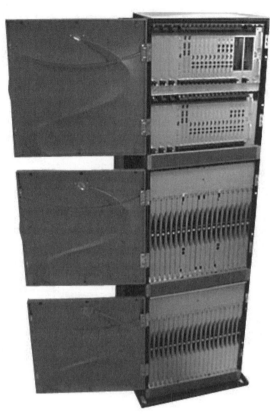

FIGURE 1-13   Spatial Wireless soft-switch/media gateway Photo courtesy of
Spatial Wireless.

DRIVING TECHNOLOGIES: COMPETING AND COMPLEMENTARY

Numerous technologies fuel Wireless Internet development, some associated with the terminals and some with the network infrastructure equipment. Many of these diverse technologies are related by the wireless access method used, whether it's GPRS or CDPD for example. We cover technologies related to terminal devices and network infrastructure in this chapter:

· Access technologies
· Application protocols and languages
· IP network design and equipment

Many technologies are used on multiple platforms, so applications must be device independent to be successful. Applications cannot be customized for a dozen different protocols or operating systems. If a Wireless Internet application is to be truly mobile, it must also function ubiquitously as the user moves from one location to another. This requires IP mobility over multiple access methods. Cellular users should be able to go indoors and on to a WLAN seamlessly. Therefore, wireless LANs should not compete with cellular or PDA wireless access, rather they should complement or extend its usefulness. Some of the leading WLAN technologies also will be discussed.

CELLULAR AND PCS-BASED TECHNOLOGIES

Cellular and PCS telephones will be the predominant mobile communications technologies for the foreseeable future. This does not mean to imply that cellular is superior to other technologies, only that it is more prevalent. With nearly 1 billion users worldwide, there is no close second. These mobile technologies are currently facing a major evolution in many areas, however, including the technical and business sectors, because they provide the economical means for realization of not only mobile computing, but also many other applications ranging from financial and retail communications to remote control and signaling.

Recent technology, whether it's TDMA, CDMA, or GSM has experienced severe physical layer and user interface constraints. Wireless data transmission is limited to relatively low speeds, 19.2 Kbps or less. Cellular display applications for data are limited to text messaging through SMS or WAP. There is a wide disparity between browser applications. Some simple graphics can be accomplished with WAP applications on some phones, whereas i-Mode in Japan offers extensive graphics including color displays on PDC phones,* even though the data rates are similar. As GPRS is launched on GSM networks, users will finally be able to access data at speeds superior to traditional dial-up Internet service, approaching 115 Kbps.

With the launch of next-generation cellular systems and 3G systems, users will have greater choice for data communications. New modems will be commercially available to take advantage of higher bandwidth networks, and packet data will replace circuit-switched data. Choices will include either SMS

* PDC in Japan is very similar to North American TDMA.

or a wireless modem using packet data at higher speeds. New high-speed data capabilities will provide the platform for mobile multimedia services, access to corporate LANs, and financial transactions from a mobile terminal. The type of service will determine the best data service to use. Many applications will find SMS satisfactory even if a modem is available for high data-rate service. And those that must use analog cellular will still have CDPD.

Other devices such as PDAs and Pocket PCs as shown in Figure 2-1 may use cellular networks or proprietary networks. (Some cellular manufacturers have integrated PDAs into their phones.)

Some cellular-based products will bear little resemblance to a cellular phone at all, such as PCMCIA modems that incorporate a cellular phone without voice capabilities. In addition to standard products, custom products will be available for wireless remote data and control applications. Examples are shown in Figure 2-2.

CELLULAR DATA MODEM TECHNOLOGIES

Today's cellular technology offers several different methods for data communications—an internal modem with built-in browser, an internal modem with an external port for connection (typically RS-232 or infrared (IR) to computer, or internal SMS messaging.

Modems fall into two categories: CDPD or traffic channel. CDPD offers true packet data communications at 19.2 Kbps whereas modems using the traffic channels are limited to the maximum rate for a traffic channel (<14.4 Kbps, depending on the standard). Future 2.5G and 3G networks will differ in two distinct ways: Traffic channel rates will be higher, from 64 Kbps to 2.4 Mbps, and all data will be either packet-based or high speed circuit-switched.

FIGURE 2-1A   Cellular PDA. Photo included courtesy of Kyocera Wireless Corp.

CDPD MODEMS

Early cellular manufacturers and operators recognized the need for data communications, and the first modems were very similar to standard modems used in homes and offices.

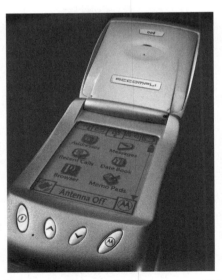

FIGURE 2-1B   Cellular PDA. Photo included courtesy of Motorola, Inc.

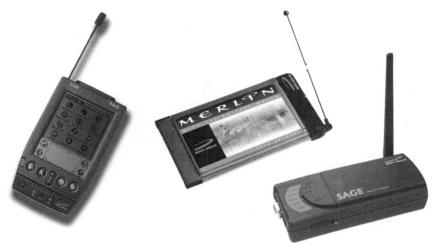

FIGURE 2-2A   Wireless PCMCIA modems. Courtesy of Novatel Wireless.

However, cellular uses a valuable, shared commodity—spectrum. (Fixed telephone lines or wires may be shared, but they can always be increased in number if necessary.) CDPD was created as a digital packet data service over an analog cellular telephone: It uses the same analog channels as voice, but with

FIGURE 2-2B   Wireless PCMCIA modem. Courtesy of Nokia Mobile Phones.

different modulation applied to the air interface. Traffic channels not being used for voice calls may be used for CDPD calls. CDPD was the first digital data application to use packet data for cellular, and it is still very much in use today by carriers such as AT&T Wireless Services.

CDPD is fully compatible with analog cellular and is colocated with AMPS systems. Therefore, the analog infrastructure, such as frequency spectrum, cell sites, towers, and antennas, can be shared. The CDPD network elements overlay parallel to the analog infrastructure (see Figure 2-3). Analog voice or analog data using an AMPS modem or digital data using a CDPD modem shares the same frequency spectrum. External modems are most common for CDPD communications, typically existing as PCMCIA cards for laptop computers, as accessories for PDA devices, or as external modems for connection to an analog phone. Some manufacturers actually include CDPD modems into their cellular telephone. This makes a 2G digital phone "Internet ready" because all TDMA and CDMA phones also include AMPS analog compatibility, and CDPD is carried on AMPS channels.

Two key design criteria were used to develop the CDPD protocol. From its inception, it was designed to use TCP/IP, the Internet protocol, making it transparent to Internet data. It was also designed to overlay an AMPS network, taking advantage of

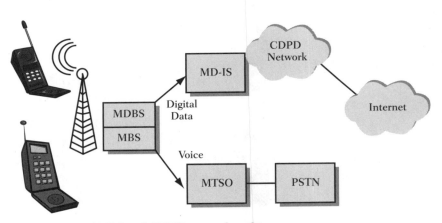

FIGURE 2-3   AMPS and CDPD network architecture.

existing infrastructure. These design goals make CDPD very attractive to any carrier that manages an AMPS network. This becomes even more important when a carrier lacks true 2.5G or 3G capabilities and finds itself at a competitive disadvantage. CDPD can make any TDMA or CDMA phone "Internet ready" with 19.2 Kbps data rates by using TCP/IP in a packet-data network instead of on a circuit-switched connection.

As of the end of the February 2000, CDPD was available in 184 markets in the United States covering 56 percent of the United States population and 60 international markets.* As 2.5G and 3G cellular networks launch, CDPD will begin to fade, but as long as there are 2G or analog phones, CDPD will retain its usefulness. And although it may compete with SMS for short text messages, it still has one advantage—CDPD works across networks with different physical layers. A CDPD modem in a laptop using CDMA can still send an email to a phone across the country, which is using TDMA.

TRAFFIC CHANNEL MODEMS

Many digital phones are advertised as Internet-ready, coming with a browser or a connectivity kit. Advertising for Wireless Internet modems or Internet-ready phones can be very deceiving, however. If the ad mentions a data rate of 19.2 Kbps, then it's CDPD. The phone may be CDMA or TDMA but the data connection is through either an internal or external CDPD modem connected to an analog channel. If a TDMA or CDMA phone has a data connectivity kit, such as a cable to connect the phone and a laptop, and it does not mention the data rate, or if the phone has an internal browser, the modem is integrated into the phone and probably uses the rate of the traffic channel, 8 Kbps for TDMA and 9.6 or 14.4 Kbps for CDMA. GSM phones have long used data capabilities built into the phones so that they connect to a laptop by cable or include built-in modems to send data in the traffic channel at 9.6 Kbps. In all of these cases except CDPD, the connection is still circuit-switched for 2G networks.

---

* Source: Wireless Data Forum.

A second type of modem is essentially a phone without voice capabilities on a PCMCIA card. There are modems of this type for every technology. They have an antenna integrated into them or are connected by a short cable to an antenna. Again, if the data rate is specified as 19.2 Kbps, it's CDPD. If the rate is not specified, it's probably using a traffic channel. Some PCMCIA modems offer a data rate of 56 Kbps and mention wireless in the same sentence. These actually combine two modems in one: a 56K landline modem and a wireless traffic channel modem.

With the launch of 2.5G and 3G networks, modems will become available having much higher data rates. They will fall into the same two categories: internal to a phone with a cable connection to a laptop or as a PCMCIA card. GPRS technology is just being launched in Europe but as with all new technologies, GPRS modems are still scarce. The United States will see next generation modems for CDMA and TDMA phones by 2002. The CDMA phones will use 1xRTT technology, and the TDMA phones will use GPRS. The CDMA standard 1xEV, with data rates up to 2.4 Mbps, will not be available until later. Products for W-CDMA will become available later in 2002 or 2003.

SHORT MESSAGE SERVICE (SMS)

If any application could be thought of as the "killer application," messaging would certainly rank high on the list. First-generation digital cellular brought new data handling capabilities to the mobile community when a new service called SMS or Short Message Service was embedded into cellular protocols. All GSM phones support SMS but not all TDMA or CDMA phones fully support SMS yet. GSM was the first protocol to use SMS so the handsets have all caught up to the feature; the other protocols are working on enabling SMS in the network as more TDMA and CDMA handsets incorporate the ability to send an SMS. Remember—all digital phones can receive text messages. Figure 2-4 illustrates an SMS network overlay.

Carriers in Europe report SMS revenues as up to 15 percent of revenues and an even greater percent of profits. Global

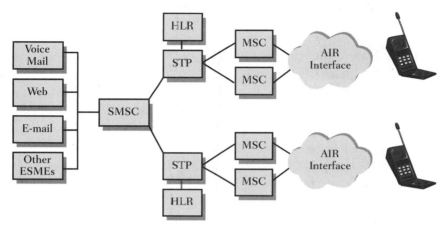

FIGURE 2-4   SMS network overlay.

SMS traffic is estimated at 15 billion messages by December 2001.* While SMS is immensely popular in Europe, it will be the first step in most carrier messaging strategies and will therefore be the first non-voice application that the majority of American consumers will experience. Although mainstream launches and promotions of SMS have taken place, widespread adoption is far from complete.

SMS teaches consumers to use wireless devices for non-voice services, and it will be the bearer for the next stage of messaging that incorporates elements other than simple text—graphics, sound, and specific formatting. Before United States customers can enjoy the same widespread usage as GSM users, however, several problems must be solved.

Not all SMS messaging is created equal. The number of characters that can be sent using SMS varies by protocol and carrier. Typically GSM sends 160 characters, TDMA sends 150 characters, and CDMA can handle up to about 200. Some phones only receive and cannot send SMS. Addressing and interconnectivity for SMS is a major challenge for substantial growth of SMS traffic. Users of CDMA cannot send SMS to TDMA users.

---

* Source: EMC World Cellular Database.

A second problem area is billing. Billing impacts the adoption of SMS because most consumers will be unsure of the need for SMS and will not have any reference point for usage. Billing can be in "buckets of messages," "per SMS," or free. Most carriers will launch SMS with a period of free SMS messages before moving to the primary offer of a bucket of SMS messages (200–800) for $4 to 8 per month.

Speed and latency offer another potential problem area in the United States. This was a problem in Europe six years ago, but because SMS is a mature technology in Europe, latency has been "fine-tuned" out. A typical SMS message is very fast— less than 5 seconds from send to receive. At times in the United States, however, the SMS traffic is so heavy (holidays, etc.) that the delay is measured in hours not seconds. SMS quickly loses value as latency increases. Carriers control latency by adding processing power to the Short Message Service Center (SMSC).

Every technology-based service such as cellular or the Internet constantly evolves into something different and, hopefully, better. SMS is no different; it will migrate to newer versions such as Smart Messaging, Instant Messaging, Multimedia Messaging, and Enhanced SMS (EMS or E-SMS).

SMS is characterized by out-of-band packet delivery and low-bandwidth message transfer, which results in a highly efficient means for transmitting short bursts of data. Initial applications of SMS focused on eliminating alphanumeric pagers by permitting two-way general-purpose messaging and notification services, primarily for voice mail. As technology and networks evolved, a variety of services has been introduced, including email, fax, and paging integration, interactive banking, information services such as stock quotes, and integration with Internet-based applications. Wireless data applications include downloading of subscriber identity module (SIM) cards for activation, debit, profile-editing purposes, wireless points of sale (POS), and other field-service applications such as automatic meter reading, remote sensing, and location-based services. Additionally, integration with the Internet spurred the development of Web-based messaging and other interactive applications such as instant messaging, gaming, and chatting.

Clearly, mobile messaging is a valuable application that is gaining popularity in both the business and consumer sectors. Mobile messaging services will continue into the next-generation networks, and multimedia messaging will emerge as more bandwidth becomes available.

General Packet Radio Service (GPRS)

General Packet Radio Service (GPRS) is a GSM Phase 2+ bearer service. It represents the first true advance in packet data service since CDPD and is the first packet data service on wireless digital networks. It is currently being launched in Europe on the GSM networks, but a common start-up problem has hampered its growth—lack of equipment! GPRS handsets are still in short supply. This is a recurring nightmare for operators of all new technologies: When WAP was introduced, there was a lack of handsets and content.

This results from the classic "chicken-and-egg" syndrome. Because GPRS handsets cost more to make in small quantities, prices to consumers are higher. With low sales figures, manufacturers produce small quantities of product. The ramp-up to higher production volumes will take time, but it will happen, of that you can be sure. GPRS will be the backbone of GSM and TDMA networks for wireless packet data communications. Radio resources are shared by all mobile stations, and GPRS parses out those resources as needed to each user because Internet browsing usually results in data communication that is transmitted in bursts rather than steady streams. This creates greater efficiency in network capacity management: Data rates as high as 115 Kbps can be achieved.

Unlike SMS messaging, GPRS was not originally a part of the GSM (or TDMA) network. For this reason, some new network elements must be introduced to the GSM architecture, and some mobility management functions must be modified as shown in Figure 2-5. Unlike CDPD, however, GPRS provides a data overlay within the standard GSM infrastructure by

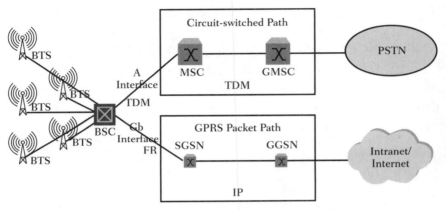

FIGURE 2-5 GPRS network overlay.

adding these additional elements. Packet data through a GPRS network does not use any circuit-switched network resources.

One of these additional network elements is called the Gateway GPRS Support Node (GGSN). Essentially, this is a packet router with some mobility management functions. It connects to the GSM network and the external packet node network through standard interfaces.

The second new element is very similar in function except that it connects directly to the Base Station Controller (BSC). The Serving GPRS Support Node (SGSN) is responsible for handling packet data to and from the mobile unit.

BT Cellnet began offering GPRS network access to mobile phone users in Europe in 2001. Recent tests of those services, however, have not impressed many customers. Actual data rates have not matched expectations, but it is a new service and there will be a great deal of "fine tuning" to the system over the next few months.

When GPRS fulfills its promise of higher data rates, many new applications will be possible over GSM and TDMA networks. GPRS will fully enable mobile Internet applications similar to Web browsing on a desktop computer. Applications will include file transfer, Web surfing, and of course, email with attachments.

As with any new technology, GPRS does have some negative impact on a network. Not only are data resources shared, they are shared with voice resources—for any given cell site, channels must be divided between voice users and GPRS users. If all voice channels are in use and file transfers are taking place on all packet-data channels, there is no more capacity for that particular cell site until someone stops using some of the resources. Dynamic allocation of resources can only do so much.

APPLICATION PROGRAMMING LANGUAGES AND PROTOCOLS

Many application protocols exist for wireless devices, such as the most widely discussed WAP and i-Mode. However, other application platforms have also achieved popular support, such as J2ME and Symbian. The strange fact is that these do not really compete with WAP or i-Mode; they can be integrated into them. We discuss each protocol briefly here.

A WORD ABOUT MARK-UP LANGUAGES

Hypertext Markup Language (HTML) was not the first descriptive text language to be used. Computer programmers have long used formatting codes, control codes or macros in software to direct document formatting. By the 1960s, generic coding began with descriptive tags rather than cryptic names. One example would be *heading* instead of *format-17*. Many people were working on similar ideas during the 1960s including scientists at IBM. A Generalized Markup Language (GML) was developed as a means of allowing the text editing, formatting, and information retrieval subsystems to share documents. GML introduced the concept of a formally-defined document type with an explicit nested element structure. Of course, GML was implemented for the mainframe computers circa late 1960s. At that time IBM was the world's second largest publisher and they produced over 90 percent of their documents with GML. Over the next few years, several new concepts were developed such as short references, concurrent document types, and link processes.

During the 1970s the American National Standards Institute (ANSI) established a committee to develop a standardized markup language. This became the Standardized General Markup Language (SGML) which was eventually adopted in 1986 by the International Standards Organization (ISO). SGML offers a detailed system for marking up documents so that their appearance is independent of specific software applications. It is a stable and well-defined meta-language that allows other markup languages to be created. SGML is very powerful and flexible due to the many options included. Early adopters of SGML were the U.S. Internal Revenue Service (IRS) and the US Department of Defense.

However, it soon became apparent that SGML's sophistication was unsuitable for quick and easy Web publishing. A simplified markup language was needed so that anyone could learn it quickly. A result was the Hypertext Markup Language (HTML), which is basically one specific SGML document type, or Document Type Definition (DTD). Early Web browsers supported HTML and it quickly became the de-facto language of the burgeoning Web and was in large part, a significant reason for the rapid growth of the Internet's popularity.

As good as HTML is, there are still problems with it. In many cases it is too simple. It served the purpose in the early days of the Web when almost everything was text-based documents but ran out of horsepower when Web authors started using multimedia and advanced page designs. Image maps (images with embedded hyperlinks), text attributes, tables, frames, and dynamic pages all added complexity. Competition among browser developers guaranteed incompatibilities with proprietary features or solutions to the same feature. Over the years Microsoft has added tags that work only in Internet Explorer, and Netscape added tags that work only in Navigator and guess what: the Web author is caught in the middle! Standards were attempted but never really got full support industry wide. The biggest problem is that HTML is not extensible. This gave way to Java and JavaScript and Active Server Pages. Each new addition to HTML such as these and Cascading Style Sheets (CSS) add flexibility in Web designs but these are really just patches to mask the problem—no standard extensibility. It is ironic that HTML grew out of SGML which is fully extensible.

As extensible as SGML is, it is also extremely complex and time consuming to customize a set of documents. A new approach was needed to bridge the gap between SGML and HTML. The answer is Extensible Markup Language (XML), a proposal in late 1996 to the World Wide Web Consortium (W3C). XML was designed with the power of SGML, avoiding the complexity. HTML is merely one SGML document type, XML is a new meta-language, a simplified version of the parent language itself. Yet, XML has the power to define other markup languages.

XML BASICS.   XML is a markup language for documents containing structured information. The information includes content and instructions for using the content. The content is flexible. It can be text, graphics or a table of information. XML represents the specification to define a standard way to add markup to documents. A document can be the traditional text with or without graphics but it can also exist in anyone of a number of other forms such as mathematical equations, server APIs, vector graphics or almost anything else.

While languages such as WML or HTML define tag semantics and the tag set, XML is really a meta-language for describing markup languages. XML specifies neither semantics nor a tag set. In HTML an <h1> is always a first level heading and the tag <cell.serial.number> is meaningless. XML provides a facility to define tags and the structural relationships between them. The semantics of an XML document will be defined in one of two ways: the applications that process them or by stylesheets. An example of XML is shown in Figure 2-6.

XML documents are composed of markup and content. There are six kinds of markup:

- Document type declarations
- Elements
- Entity references
- Comments
- Processing instructions
- Marked sections

```
<?xml version="1.0"?>

<anotherhelloworld>

<speaker1>Say <quote>Hello World.</quote></speaker1>

<speaker2><quote>Hello World.</quote></speaker2>

<yawn/>

</anotherhelloworld>
```

FIGURE 2-6   A simple XML document.

Each of these types of markup are clearly defined in the XML standard, Our intention is not to give a detail explanation on XML, only illustrate the similarities and differences to other markup languages.

XML allows groups of people or organizations to create their own customized markup applications for exchanging information in their domain. Examples include music, chemistry, electronics, finance, scuba diving, petroleum geology, linguistics, cooking, knitting, history, engineering, mathematics, or baseball card collecting.

COMMON GROUND—XHTML

The HTML language today is ubiquitous in many millions of Internet documents but browsers today pay a heavy price for the need to parsers need to accommodate bad syntax and authoring practices. It is be very difficult to create a "light browser" which is important in wireless applications. This of course, has spawned the need for other languages such as Wireless Markup Language (WML) and c-HTML or Compact HTML. Clearly a better solution needs to be found and it is generally believed that XML will be the leading contender. In 1998 the W3C published a draft document entitled "XHTML 1.0." This is an attempt to merge HTML 4.0 and XML.

Current thinking is that other markup languages, WML, HTML and c-HTML, will "circle the wagons" around XHTML.

However, Web authors will need stricter discipline and training to use these tools. The upside will better documents, more compatibility and the ability to streamline browsers for both the wired and Wireless Internet.

Note: A free little program that converts an HTML page to XHTML for you, along with correcting many common authoring mistakes may be found at www.w3.org/People/Raggett/tidy/.

Changing from other markup languages to XML or XHTML more specifically will enhance and extend the utility of Internet to new applications. However, XML and XHTML are new, so do not expect current browsers to work flawlessly when usage of these languages begins on Internet sites.

Some of the generic parts of XML such as parsing, tree management, searching, and formatting are being combined into libraries to make it easier for developers. Some applications can use languages like Java to develop plug-ins for generic browsers. Microsoft Internet Explorer 5.5 handles XML but through an HTML model. Netscape 6 also includes some XML support.

WIRELESS APPLICATION PROTOCOL (WAP)

WAP is an application protocol for providing communications and connectivity to the Internet using digital mobile phones, pagers, personal digital assistants, and other wireless devices. It allows a user to navigate a text menu on his phone and click on a menu item to select the next action.

Unlike other standards, however, the WAP standard was not created by an industry consortium like the European Telecommunications Standard Institute (ETSI) or Telecommunications Industry Association (TIA). It began as the idea of a small group of men with a vision at a company called Libris. Out of that vision, a standard was born. Wireless Application Protocol is based on a client–server architecture much like that used on computers. Because of the complexity and size of the cellular phone, a decision was made to place the server on the back end.

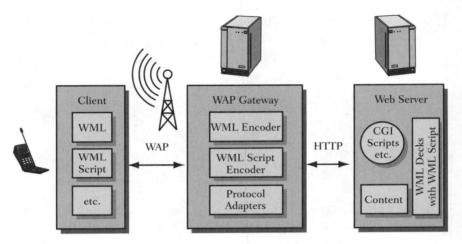

FIGURE 2-7   WAP programming model.

The programming model for WAP as in Figure 2-7 consists of a WAP Gateway and a Wireless Application Environment (WAE) User Agent in the wireless device. Data flows from the wireless device to the Gateway in encoded requests. These are forwarded on to the Internet. The response, in the form of Internet content, is returned to the Gateway where it is encoded and forwarded to the User Agent in the wireless device. Applications are written in Wireless Mark-up Language (WML) and can reside on the WAP Gateway or an application server. The Gateway converts protocols between message origins and destinations. WAP also includes WMLScript and Wireless Telephone Application Interface (WTAI). The script provides a simple, yet fully functional programming language to run within WML applications. WTAI is an interface to the telephone functions of the device. For example, if you send a request to directory assistance for a new restaurant that you want to try, the phone number is sent to your user agent; by highlighting and clicking Enter, the number is dialed automatically.

Libris developed a computer-generated prototype in early 1995 that was used to demonstrate WAP principles to potential customers and investors. The key to getting WAP started was buy-in from carriers and phone manufacturers. The Big Three phone manufacturers—Motorola, Ericsson, and

Nokia—had to be convinced to drop proprietary plans and back a standard-based proposal. Through relationships with some key carriers like GTE and AT&T Wireless Services, Libris convinced manufacturers to support the WAP initiative. The only way for WAP to succeed was as an open standard: It could not remain the sole property of its inventor and achieve widespread usage. With a potential market of more than 1 billion cellular phones by 2002, it was in the best interest of all manufacturers to agree to standardization. By mid-1996 Libris changed its name to Unwired Planet.

The first commercial server and browser software was released in June 1996. During the next twelve months, Unwired Planet collaborated with Ericsson, Motorola, and Nokia to create the first WAP architecture, and published it on the Internet in September 1997. Figure 2-8 shows some of the WAP milestones.

A single focal point for WAP development, the WAP Forum, was created in January 1998. The WAP Forum was created as an open membership organization to promote Wireless Application Protocol, write and issue new revisions to the WAP standard, and to develop a quality-assurance function to verify compatibility issues between members. In May 1998, the WAP Forum released the first, open-version WAP 1.0 specifications.

WAP, however, has not been an instant success. The biggest complaint about WAP is that it required the Internet community to rewrite sites to support WAP. WAP proponents are quick to point out, however, that the mobile phone is an entirely different user interface compared to the typical Internet application. The display is much smaller and the keypad lacks the full

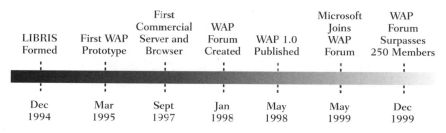

FIGURE 2-8   WAP milestones.

alphanumeric keyboard that computer users enjoy. It should not be expected that Internet content could appear the same on a phone as on a computer monitor. Internet content must be optimized for presentation to the phone and that presented the next big challenge—getting application developers to commit to WAP development.

Another problem common to WAP is inconsistent adherence to the WAP standard—every phone seems to behave differently and support different features. This occurs not only between manufacturers, but between different phones of the same manufacturer. One phone may center-justify text and the next ignores all text justification. Displays vary in size and the number of characters or lines, creating a different look to every phone. A common mistake in WAP design is to include too many nested menus for navigation. Better user interfaces and larger displays and easier-to-use keypads must be created to improve adherence to the standard.

To prevent wars between rivals, proponents of WAP have agreed to make XML (Extensible Markup Language) WAP-compliant. The expectation is that WAP will merge with another popular application protocol, i-Mode, to become Extensible Hypertext Markup Language (XHTML). An example of WML is illustrated in Figure 2-9.

Wireless Application Protocol is a major player in the microbrowser market. It was first launched in Europe, where consumers already enjoyed SMS on GSM networks. Competition was tough in Europe, however—although WAP

```
<?xml version="1.0"?>
<!DOCTYPE wml PUBLIC
  "-//WAPFORUM//DTDWML 1.1//EN"
  "http://www.wapforum.org/DTA/wml_1.1.xml">

<wml>
  <card title_"Hello World Example">
    <p>Hello, World!</p>
  </card>
</wml>
```

FIGURE 2-9   WML code "Hello World."

was launched, there was a shortage of WAP-enabled phones. Content was lacking and no two phones worked or displayed WAP the same. A running joke in Europe described WAP as standing for "Where Are the Phones?" Despite everything, Europe is estimated to have over 16 million WAP subscribers.

In the United States, the situation is even worse. Not only is there a shortage of WAP phones and applications, but TDMA carriers never deployed WAP gateways because of delays in getting licensing from manufacturers. Meanwhile, in Japan, WAP is deployed on CDMA phones that compete directly with i-Mode. Japanese sources estimate that there are about 6 million WAP users compared with over 20 million i-Mode subscribers.

One important distinction should be made between today's WAP phones and next-generation phones. Until recently the current generation of standards could only support circuit-switched connections. Packet data on digital networks did not exist. Just like CDPD, these phones remained on a dedicated channel assignment as the user browsed the WAP application. The launching of GPRS in the GSM world enabled the first packet-data network.

Wireless Application Protocol will continue to exist but the big question is "in what form?" Many industry experts believe that the window of opportunity is closing on WAP. The WAP Forum is a large and strong industry body, however, and a great deal of investment has already been made in WAP. The industry will continue to launch the technology wherever there is a possibility to create new revenue streams for all.

i-Mode (Compact HTML or c-HTML)

NTT DoCoMo first introduced i-Mode in Japan in February 1999. Since then it has been immensely successful. There were more than 20 million users by March 2001. Despite the fact that i-Mode does not yet exist outside of Japan, i-Mode accounts for over 60 percent of the world's mobile Internet users! There are over 1,500 official Web sites and over 40,000 independent sites. In fact, i-Mode Web sites may be built anywhere in the world and many times are included with HTML Web server sites by placing the code in a subdirectory.

i-Mode is a client–server protocol similar to WAP and illustrated in Figure 2-10. It allows users to navigate a series of menus on their phone display in order to access Internet content on i-Mode sites.

One very important comment must be included here about i-Mode. NTT DoCoMo created i-Mode as a proprietary protocol without international standards' body cooperation. The result is a tightly controlled application environment that NTT DoCoMo designs and markets to their best advantage. It should also be noted that DoCoMo is one of the first carriers to offer usage-based billing for i-Mode. Because i-Mode is packet-based, they charge for the actual volume of data transmitted.

i-Mode phones in Japan operate on the Personal Digital Cellular (PDC) network, which is very similar to the North American TDMA network. Therefore, the data rate is limited to 9.6 Kbps. In the future, when W-CDMA is launched in Japan, much higher data rates will become available.

Although i-Mode can only be used on phones inside Japan, NTT DoCoMo clearly intends to make it a world standard for Web browsing on wireless devices. Their recent investment in AT&T Wireless, the agreement giving AT&T access to i-Mode technology, and the purchase of an ISP—Verio Communications—demonstrates that they will be very proactive in advancing i-Mode throughout the world. They have also taken a 15 percent stake in Dutch KPN Mobile and claimed 20 percent of Hutchison 3G.

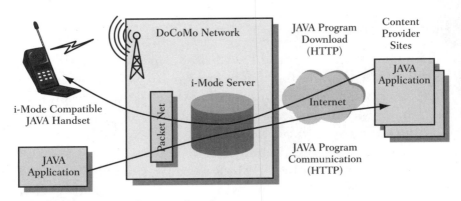

FIGURE 2-10   i-Mode network architecture.

Unlike WAP, which uses WML as its markup language, i-Mode services are built using i-Mode-compatible HTML, referred to as compact HTML or c-HTML. On the surface, both look very similar and in fact, both markup languages have common elements. An example of i-Mode code is shown in Figure 2-11. It looks just like HTML to the seasoned Web programmer with the exception of the accesskey command. Unlike WAP, there is no card deck, but i-Mode does support color displays.

DoCoMo based i-Mode on a subset of HTML 2.0, HTML 3.2, and HTML 4.0 specifications with i-Mode specific extensions. These extensions include tags that have special meaning on a cellular phone such as the tel: tag, which is used to hyperlink a telephone number. By clicking on the link, users can place a call to the number.

Just as in WAP, some i-Mode phones support more than the basic standard protocol. Features that require cursor movement in two-dimensions, such as tables, are not supported.

```
<html>

<head>
<META http-equiv="Content-Type" content="text/html; charset=utf-8">
<META name="CHTML" content="yes">
<META name="description" content="sample cHTML document">
<title>Sample cHTML document</title>
</head>

<body text="#000000">
<center>Links</center>
<hr>
Select an option<br>
<a href="message.chtml" accesskey="1">1</A>Messages<BR>
<a href="mail.chtml" accesskey="2">2</A>Mail<BR>
<hr>
<center>
<a href="admin@chtmldemo">email: admin@chtmldemo </a><br>
</center>
</body>

</html>
```

FIGURE 2-11   i-Mode code example.

Included in i-Mode are predefined symbols that can be called using embedded HTML text as a 6-byte decimal code or embedded Shift-JIS 2-byte hex code into HTML. Graphic characters that are part of the standard S-IS character set can also be used. GIF images including animated images can also be displayed on i-Mode phones.

English language programmers should also be aware that English words do not break correctly at a line ending so that programmers must specify tags at word breaks, or words will be wrapped to the next line. While there are few tools in English for i-Mode programmers, at least one tool is available—Adobe GoLive 5.0 includes i-Mode and WAP functionality.

Japan's i-Mode service offers more affordable access rates (averaging of $70 a month for voice services and $17 for i-Mode transmission fees), has more robust content than WAP, and is faster. However, NTT DoCoMo's i-Mode will be difficult to make available on other service carriers' networks. The other side of the coin is that Web sites are historically HTML-based, and WAP is not compatible with HTML. NTT DoCoMo is a member of the WAP Forum and hopefully the two proponents will converge on a common standard, HTML 4.0 or XHTML. Mobile phone and browser manufacturers are going to support XHTML in next-generation phones. At least two industry leaders, Nokia and Openwave, have announced their intent to support XHTML mobile browsers by 2002.

J2ME

Java Version 2 Mobile Edition (J2ME) is a compact version of Sun's Java technology targeted for embedded consumer electronics. J2ME includes a set of developer tools and supplies to be used in mobile applications. As with other Java platforms, key advantages can be realized using J2ME: Consistency across products, code portability, reliable network delivery, and leveraging of the Java programming language. It is also upwardly scalable to work with other Java platforms.

Sun Microsystems is targeting J2ME for devices with limited hardware resources such as PDAs, cellular phones, and other consumer electronics and embedded devices with as little as

128 Kb of RAM. J2ME actually consists of a set of profiles and configurations as in Figure 2-12. Each profile is defined for a particular type of device—cell phones, PDAs, smart appliances, and others. The profile specifies the classes, methods, and configuration needed. The configuration includes a minimum set of class libraries required for the particular type of device and a specification of a Java Virtual Machine (JVM) required to support the device.

J2ME technology enables wireless device manufacturers, service providers, and content providers to perform rapid, cost effective development of new features and applications. The latest release includes emulation tools for Palm OS, RIM, and others. Also included is support for Secure Hypertext Transfer Protocol (S-HTTP).

Using J2ME, developers can write applications such as a WAP browser for a cellular phone or a PDA. J2ME allows for third-party developers to create applications that run on manufacturers' devices without providing proprietary information or special application programming interfaces and documentation.

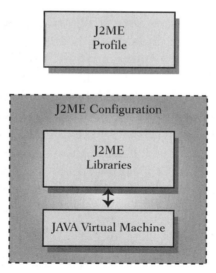

FIGURE 2-12   J2ME profile and configuration.

SYMBIAN

Symbian is a generic operating system developed for wireless communications devices. Included is a set of common core Application Programming Interfaces (APIs). Symbian makes a distinction between the generic technology used in any Wireless Information Device (WID) and the Graphical User Interface (GUI) tailored for a particular design. A generic design is the heart of Symbian. Technology is shared between all reference designs.

The operating system includes a kernel; middleware for communications, data management, and graphics; the lower levels of the GUI framework; and application engines. Products such Ericsson's R380, Nokia's 9210 Communicator, and Psion's Series 5mx were created with very little modification to the Symbian OS. Two reference designs were created, one for information-centric products such as the Nokia Communicator and another for voice-centric products such as "smart phones," mobile handsets with built-in browsers.

Symbian's latest version includes support for GPRS and WAP 1.2.1.; tablet or keyboard entry; and application development using C++, J2ME, WAP, or HTML. Support for protocols such as TCP/IP, WAP, GSM, GPRS, Bluetooth, IrDA, or RS-232 is built into the operating system. Many other features such as security, font and text formatting, and a rich suite of application engines are included.

Symbian uses a generic technology for the specific requirements of wireless devices. The requirements are tailored to use device resources efficiently, and to be reliable and adaptable to device needs. The architecture is illustrated in Figure 2-13.

Manufacturers may use Symbian reference designs and operating system to reduce time-to-market for new product development. Support for almost every conceivable wireless device is included, and any application development platform can be overlaid on the operating system, such as WAP or J2ME.

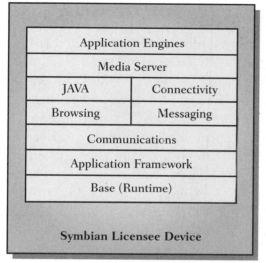

FIGURE 2-13 Symbian generic architecture.

WinCE

Windows CE is the modular real-time embedded operating system from Microsoft. It is a scaled-down, multitasking, multi-threading look-alike of the popular Windows Operating System designed for 32-bit devices or information appliances. However, it does not require an Intel x86 family microprocessor. Several manufacturers, including Compaq, are currently marketing products using this operating system. Although it looks similar to Windows, it does not necessarily run all Windows applications without modifications. When applications are run on devices other than PCs, some Windows applications will not run at all but this does not mean that it does not serve as a useful tool.

The most important thing to remember about WinCE is that the interfaces run a shell that is very similar to the standard Windows shell, with the same windowing look-and-feel and a taskbar at the bottom that can be used to launch and control applications. Unlike PDAs (like Palm), where a new interface must be learned, WinCE has a familiar "look and feel." One major criticism, however, haunts WinCE—the large amount of memory required.

PDA AND POCKET PC TECHNOLOGIES

Wireless devices are not always cellular phones. In fact, they do not even need voice capabilities. Does you home PC have a telephone receiver hanging on it? Of course not: The computer's primary function is to do meaningful data crunching!

Personal digital assistants and pocket PCs were created to provide a certain level of support for the mobile user, which could only previously be available on a laptop PC, but smaller and lighter. They are not primarily voice communications devices—they are for storage, retrieval, and manipulation of data. Some PDAs and pocket PCs may attach to a cellular phone or other type of wireless modem, but only for the wireless connectivity of data communications. At least one manufacturer's cellular phone has a PDA built-in. It is immaterial whether we choose to call it a phone with PDA or a PDA with a phone built into it.

There is an entire class of devices devoted to wireless mobile data. Some PDAs are primarily email devices. Others are more generic and can run many different applications, such as the Palm Pilot or Visor. All, however, fall into two categories: They either use public cellular for communications or they use other public or private networks. They all share the same limitations for data transmission, and they are limited to the current state of the art of the physical access channel. Except for Ricochet, this means 19.2 Kbps or less. When they are upgraded to next-generation cellular, data rates will increase.

One very interesting thing to watch will be the migration or upgrade paths for the noncellular networks when competition for higher data rates begins. Most are using 19.2 Kbps for wireless PDA and Pocket PC communications.

RIM AND BLACKBERRY

Research in Motion makes RIM and Blackberry wireless handheld models for a variety of applications and personal preferences, including an OEM version. Some models are primarily for Internet email, whereas other models include the functionality to run application software such as calendars, address books, and much more. The devices fit in the palm of a user's hand and are lightweight. Wireless communications is through one of two different networks, Mobitex or DataTAC (Ardis Network).

Both Mobitex and DataTAC are packet-switched, narrowband PCS networks designed for wide-area wireless data communication at a data rate of 19.2 Kbps. They provide always-on connections with extensive coverage and in-building penetration, seamless roaming, fast messaging, high reliability, and advanced messaging services. Dialing up to retrieve emails is unnecessary.

PALM

Palm offers several Wireless Internet options. Their integrated wireless solution is the Palm VIIx with service provided by Mobitex. Other Palm products can add a modem or connect to a cellular phone. One wireless service used by Palm is OmniSky, which offers CDPD at 19.2 Kbps. This, of course, operates over an analog cellular network. Because analog cellular coverage is very ubiquitous in North America, this provides broad user coverage when traveling. It also carries all of the advantages of CDPD, like packet data transmission and always-on mode.

Downloading Web pages with Palm products involves a little trick called *Web clipping*. The user requests a particular piece of information in a query, and the page is returned minus a lot of the extraneous details. A server gateway is used to optimize the content for displaying on the handheld PDA.

HANDSPRING

Handspring products also use CDPD modems, just like Palm, for wireless communications. In addition to CDPD modems, Handspring offers an IEEE 802.11b wireless LAN option, which offers connection speeds up to 11 Mbps in a WLAN environment. A third wireless option for Handspring products is a wireless modem connecting to the Glenayre messaging network through SkyTel service. This is basically email service through two-way paging and not suitable for Wireless Internet.

HP

Wireless Internet solutions for HP products, PDAs, and Pocket PCs are CDPD modems sold by third parties.

COMPAQ

Compaq Pocket PC products have the greatest number of options for Wireless Internet of all PDAs and Pocket PCs. Through third-party suppliers, a user may select PCMCIA modems using CDPD, Ricochet, or CDMA networks.

PROPRIETARY NETWORKS

ARDIS (DATATAC)

Advanced Radio Data Information Services (Ardis) sometimes referred to as the DataTAC network, is a two-way radio service that is based on Motorola's RD-LAP technology. It was originally created and jointly owned by Motorola and IBM to serve IBM field technicians. Coverage includes about 90 percent of the urban business population in the U.S with more than 400 metropolitan areas in the United States, Puerto Rico, and the Virgin Islands.

The network is based on packet-data transfer using data rates as high as 19.2 Kbps. Some areas may not offer rates as high as 19.2 Kbps if they are not enhanced with RD-LAP. Frequencies and protocols are proprietary to Motorola. Modulation at the physical layer is Gaussian Frequency Shift Keying (GFSK).

RICOCHET

Ricochet is the only wireless packet data network today that was built from the ground up to handle high speed data, up to 128 Kbps. Modems are available from third-party sources that allow connection to laptop computers by at least one Pocket PC. The network has about 41,000 customers in 15 markets as of July 2001 but the parent company, Metricom, has filed for Chapter 11 in bankruptcy court, so anything could happen to Ricochet. On August 8, 2001, Metricom ceased operations of Ricochet but the company is negotiating with third parties to buy the Ricochet network. If a third party can buy the network at a really good price, then Ricochet may continue in some form. As of the publishing of this book, the outcome is unknown. The high costs of building out a nationwide network far exceeded the revenue, and current economic conditions make financing very difficult.

The Ricochet radio network utilizes several elements including microcell radios and wired access points. This provides true Wireless Internet access to information—regardless of where it resides either on the Web or the intranet, in an email message or a video clip.

In many ways the Ricochet network functions similarly to a wireless Ethernet network. Acquisition is a necessary first step for each radio on the network. The user radio, when first turned on, must locate neighboring radios and Ricochet modems by sending out synchronization packets. On acknowledgment from neighboring radios, it must then get the authorization from the name server. Only then does it join the network.

The network operates in two Industrial, Scientific, Medical (ISM) bands of regulated, unlicensed spectrum, the 900 MHz band and the 2.4 GHz band, in addition to the licensed 2.3 GHz Wireless Communications Systems (WCS) spectrum. The physical layer uses Frequency Hopping Spread Spectrum (FHSS) technology.

Mobitex or RAM Mobile Data

Mobitex enjoys wide acceptance as a global standard for wireless data networks. This technology was originally developed by Swedish Telecom as a private network similar in purpose to the Ardis network. The network became commercial in 1986. Since then, many networks have been deployed in Europe, the United States, and Australia. The frequency varies by country but the United States and Canada mainly use 900 MHz. In the United

States, Mobitex is operated by RAM Mobile Data, a subsidiary of
Bell South. There are over 1,200 base stations installed nationwide with service in more than 7,700 cities and towns, covering
approximately 93 percent of America's urban business population.

Mobitex technology offers six distinguishing features that
other networks lack:

- Transparent, seamless roaming
- Store-and-forward
- Dependability (99.99 percent)
- Interoperability and connectivity options
- Capacity to support millions of subscribers
- Security second to none

Channels use 12.5 kHz bandwidth and support a data rate
of 8 Kbps. The network operates in the United States at 935
MHz to 940 MHz for the downlink (base to mobile) and 896
MHz to 901 MHz for the uplink (mobile to base).

OMNISKY

OmniSky's Wireless Internet service uses the CDPD packet-
based network, encompassing over 172 million people. The
first service began in May 2000. Data rates are 19.2 Kbps and
users are offered a flat-rate monthly fee for unlimited service.

Much of Omnisky's success can be attributed to the partners it has chosen to work with: Palm, Handspring, HP, and
Compaq. However, Omnisky should be feeling the competition
when cellular begins next-generation service.

WIRELESS LANS AND PERSONAL AREA NETWORKS

The Wireless Internet is not just wireless communications across
town or the country. It is also local—sometimes in a home or
office building. Wireless LANs are just becoming popular with

economically priced wireless Ethernet equipment. Standards such as IEEE 802.11, HiperLAN2, and Home RF are leading the way to untethered communications in-building or outside over small areas. Another important development is the Personal Area Network, also known as Bluetooth. Let's take a look at each of these to see how they further promote Wireless Internet sessions.

BLUETOOTH

Bluetooth is a low-cost, low-power, short-range radio link for mobile devices and for WAN/LAN access points. It operates in the ISM band. The Bluetooth standard was created primarily to replace serial cables between computers and printers or other peripherals. Speed and reliability were key considerations. Bluetooth is capable of both voice and data communications at speeds up to about 70 Kbps.

Bluetooth technology is an enabling technology for the Wireless Internet and the mobile user. It can be an Internet bridge between a mobile device and a wireless access point in an ad-hoc network, as are other WLAN technologies such as 802.11 or Home RF. However, some features of Bluetooth are unique to it and not available in other WLAN technologies. Bluetooth actually creates a Personal Area Network. It is small enough to be embedded in everyday devices such as headsets or microphones. It can be embedded in a PDA and automatically synchronize a computer to a PDA. Bluetooth can also download a file or picture received on a Wireless Internet phone to a printer or a PDA or computer.

Applications for Bluetooth wireless technology come from no only the telecom industry but also from the computer, home entertainment, automotive, health care, automation, and toys industries. What good is a wireless Internet session if you must constantly connect to wired network to print? Bluetooth uses a low-cost short-range radio link or bridge between Bluetooth enabled devices. Computers, phones, printers, wireless headsets, and microphones can all communicate with each other without wires being dragged about. Bluetooth started as an idea in 1994 at Ericsson. Today, the Bluetooth SIG boasts almost 2,500 members with nearly every major communications company represented.

Bluetooth computer and telecom consumer products will appear in late 2001 or early 2002. Products in other industry sectors will become available later in 2002.

The Bluetooth Specification addresses two ranges: short (around 10 m) and medium (around 100 m). The radio link is capable of voice or data transmission to a maximum capacity of 720 Kbps per channel. The radio spectrum used is in the unlicensed ISM band at 2.4 GHz. Modulation is Frequency Hopping Spread Spectrum (FHSS).

Because Bluetooth encompasses many applications, there is no single competitive technology. Infrared is a competitor in some cases but it requires line of sight, whereas wireless LANs have much greater range. Perhaps the closest competitor is Home RF but it too is more a wireless LAN than a personal area network.

IEEE 802.11

Wireless Ethernet is IEEE 802.11b today, the IEEE standard for wireless LAN's. IEEE 802.11b operates in the ISM band at 11 Mbps. However, several new versions of the standard is being developed, 802.11a, which supports data rates of up to 54 Mbps, and operates in the 5-GHz UNII (Unlicensed National Information Infrastructure) band. Another version 802.11g is currently being developed which will support up to 20+ Mbps. Table 2-1 summarizes the different versions of 802.11 and includes HiperLAN2 for comparison. It should also be noted that the IEEE is working on 802.11e, a standard that spans home and business environments with QoS and multimedia support while maintaining full backward compatibility with 802.11b and 802.11a. This version will support voice and include a higher level of security than 802.11b. The release date for the standard is unclear at this time.

The IEEE802.11b specification was finalized in 1999 and quickly adopted by many companies. However, it was just as quickly discovered that there are two problems: the security is weak and the theoretical transmission speeds of 11 Mbps falls short—real world speed is only about 7 Mbps.

TABLE 2-1 IEEE802.11 Versions

VERSION	802.11	802.11B	802.11A	802.11G	HIPERLAN2
Band	2.4 GHz	2.4 GHz	5 GHz	2.4 Ghz	5 GHz
Data Rate	2.4 GHz	11 Mbps	54 Mbps	20+ Mbps	45 Mbps
Standard Body	IEEE	IEEE	IEEE	IEEE	ETSI
Proponents	All manufacturers	All manufacturers	U.S. manufacturers and Nokia	Unclear at present	European manufacturers
Release Date	1997	1999	4Q2001 (maybe)	2002?	2002

Figure 2-14 depicts the OSI software model for 802.11. It closely resembles the approach taken for the HiperLAN2.

FEATURES.   IEEE802.11 also supports infrastructure networks and ad hoc networks. One very important characteristic of 802.11 is that the data rate will be automatically decreased as signal deteriorates between the access point and the stations. While 802.11b does include a security mechanism, it has been discovered to be weak. It also supports station roaming between access points.

UNLICENSED SPECTRUM USAGE FOR WLAN.   The Federal Communications Commission (FCC) specifies the rules for operating in the unlicensed 2.4 GHz spectrum. The largest governing concern is harmful interference with authorized services and must work around any interference that may be received from phones, microwaves or other RF devices.

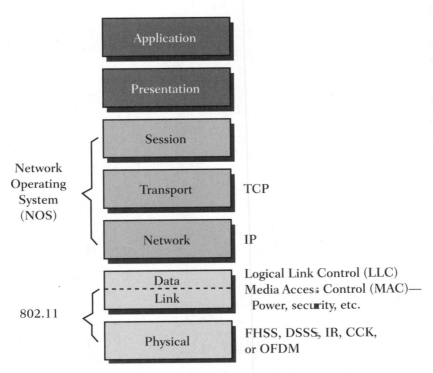

FIGURE 2-14   IEEE802.11 approach to OSI Model.

The FCC mandates that a device must operate in one of two ways in the 2.4 GHz ISM band:

- *Frequency Hopping Spread Spectrum (FHSS)*. The frequency changes in a pseudo-random manner based on a predefined code.
- *Direct Sequence Spread Spectrum (DSSS)*. The data signal is broken up into sequences and transmitted to the receiver, which reassembles the sequences into the data signal.

Future versions such as 802.11g may adopt OFDM if the FCC decides to support it and the industry can agree to rally behind it. However, at the time of publishing this book, these are two very big "ifs."

It is estimated that more than 7.8 million wireless LAN chipsets were produced in 2000. A similar number is expected in 2001. Sales are growing from almost $400 million in 2000 to $1.2 billion by 2005. Costs have dropped during 2001, causing widespread usage in homes and enterprise systems. However, 2002 will see the release of more Home RF and 802.11g products also. Parks Associates estimates that, while 5 percent of U.S. households currently have a PC network in place, as many as 15 percent will have one in five years. Of that, wireless networking will account for 40 percent of all those home networks.

HiperLAN and HiperLAN2

HiperLAN or more recently, HiperLAN2 are standards approved by the European Telecommunications Standards Institute (ETSI). HiperLAN2 is the most recent version. It is an interoperable standard providing high-speed, broadband connectivity for wireless LANs in corporate environments, public "hot spots" and home environments

HiperLAN2 provides a 54 Mbps data rate on the globally allocated 5.15–5.3 GHz band. It also may be used in the 17.1–17.3 GHz band in certain geographic locations. It surpasses the IEEE 802.11a standard with both greater security and traffic prioritization capabilities. HiperLAN2 also includes mechanisms for handoffs between WLANs and 3G mobile systems.

Currently several European manufacturers are implementing solutions that provide a wireless Virtual Private Network (VPN) solution for HiperLAN 2 which includes authentication and encryption. This will enable wireless mobile users to have a secure connection to their corporate networks when traveling through so called "hot spots," such as airports, hotels and conference centers.

HiperLAN2 achieves its high data rate by using a frequency multiplexing method called Orthogonal Frequency Digital Multiplexing (OFDM) with various physical layer modulation schemes as shown in Table 2-2.

TABLE 2-2   Physical Modes Defined for HiperLAN2

MODE	MODULATION	CODE RATE	PHY BIT RATE	BYTES/OFDM
1	BPSK	1/2	6 Mbps	3.0
2	BPSK	3/4	9 Mbps	4.5
3	QPSK	1/2	12 Mbps	6.0
4	QPSK	3/4	18 Mbps	9.0
5	16QAM	9/16	27 Mbps	13.5
6	16QAM	3/4	36 Mbps	18.0
7	64QAM	3/4	54 Mbps	27.0

OFDM is particularly efficient in time-dispersive environments, i.e. where the radio signals are reflected from many points such as in offices. The basic idea of OFDM is to transmit broadband, high data rate information by dividing the data into several interleaved, parallel bit streams, and let each bit stream modulate a separate subcarrier. HiperLAN2 is time-division multiplexed and connection-oriented. It can be used for point-to-point or point-to-multipoint connections. A dedicated broadcast channel is also included. Each connection can be assigned either a simple relative priority level or a specific QoS in terms of bandwidth, delay, jitter, bit error rate, etc. Hiperlan2 uses an approach for the Access Channel that differs from the OSI model but is very similar to the IEEE 802-11 standard as seen in Figure 2-15.

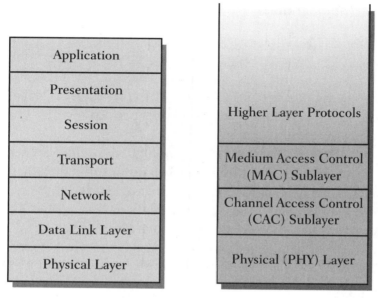

FIGURE 2-15   A comparison of OSI and HiperLAN2 access channel implementations.

HiperLAN2 was designed for short range communications, about 150 feet maximum. It is primarily meant to be used in a stationary environment but does support mobility up to 4.3 feet/second. It may be used on networks with or without infrastructure to support isochronous traffic such as audio or video with minimum latency. It can support asynchronous traffic data of 10Mbps with immediate access. HiperLAN2 is also compatible with ATM.

Radio-based wireless LANs tend to exhibit randomized "bursty" traffic patterns which can result in performance issues. Many factors have to be taken into consideration, when quality of service is to be measured. Among these are:

· Landscape topography
· Elevations that might cause shadows
· Multi-path from signal-reflection surfaces
· Signal loss through absorbing surfaces
· Quality and placement of the wireless equipment

- Number of stations
- Interference
- Etc.

These and other factors have been figured into the HiperLAN2 specification to allow for a certain level of Quality of Service guarantee.

Figure 2-16 depicts a typical topology of a HiperLAN2 network. The Mobile Terminals (MTs) communicate with one Access Point (AP) at a time over an air interface. As a user moves from one AP to the next, handoffs can take place. In an ad hoc networks, the MTs communicate directly., can also be created, but their development is still in early phase. The HIPERLAN/2 is planned to be finalized by the end of 1999.

HIPERLAN2 FEATURES.   Other than the high data rate and QoS features, HiperLAN2 includes including the following:

- Automatic frequency allocation
- Security support
- Mobility support

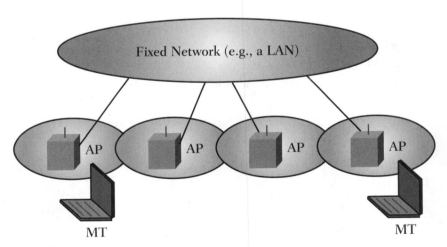

FIGURE 2-16   A HiperLAN2 network.

· Network and application independent
· Power save mode

Automatic frequency allocation is especially important because this allows for easy installation without the need for complicated frequency planning such as that required for cellular. The access points use a built-in support for automatic transmission frequency allocation.

HiperLAN2 networks also supports authentication and encryption. A handoff mechanism is managed by the mobile terminal based on received signals from each access point. Connections are maintained just in cellular (hopefully maybe even better). The HiperLAN2 network may also integrated with a variety of fixed networks.

A power save mechanism is based on mobile terminal-initiated negotiation of sleep periods. A request is made to the access point for a low power state and a specific sleep period. At the end of the sleep period, the mobile terminal searches for a wake up indicator from the access point, and in the absence of that, sleeps the next period, etc.

HOME RF

Another industry group, the Home Radio Frequency Working Group (HRFWG)—made up of members of industry leading companies such as Compaq, Ericsson, HP, IBM, Intel, Microsoft, Motorola, and others—created the Home RF Standard Specification. Home RF combines elements of 802.11 and Digital Enhanced Cordless Telecommunications (DECT) but supports only up to 2 Mbps. It is aimed at homes and small businesses.

The price of Home RF is generally less expensive than 802.11 but performance is considerably less. The devices operate in the 2.4 Ghz ISM band just as 802.11 devices do. In actuality Home RF competes more with Bluetooth than 802.11. It was designed for embedded applications in appliances and computing equipment such as printers. Only time will tell if this standard prospers.

INFRASTRUCTURE PROTOCOLS AND APPLICATIONS

H.323

H.323 defines packet standards for terminal equipment and services for multimedia communications over local and wide area networks communicating with systems connected to telephony networks such as ISDN. The initial version of this standard came from the International Telecommunications Union (ITU) in June 1996.

It defines communication over IP-based local area networks (LANs). A later version (v2), adopted in January 1998, extended it over wide area use and general-purpose IP networks. Several subprotocols are included under H.323 relating to call setup and signaling.

Four components for a multimedia communication system as shown in Figure 2-17 include terminals, gateways, gatekeepers, and multipoint control units (MCU). Gateways and gatekeepers are used in negotiation for PSTN connections, whereas MCUs enable multiparty audio and videoconferences.

One drawback of H.323 is that it is somewhat complex and inflexible. However, it is ISDN-based and relatively easy to build applications across it. For many applications, H.323 is

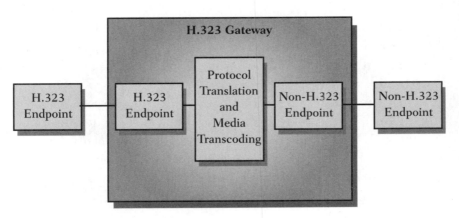

FIGURE 2-17   H.323 interface.

satisfactory, but falls short for more advanced implementations and solutions.

All things considered, the most likely scenario is that multiple protocols will be used with H.323, such as SIP for exchange between soft switches and gateways and MGCP for call setup, because H.323 is too complex and time consuming to set up a call.

MGCP/MEGACO

The Media Gateway Control Protocol (MGCP) specifies communication between call control elements and telephony gateways. It is a text-based protocol. Media gateways are telephony gateways that convert circuit-switched voice signals to data packets for multiservice packet networks. The Internet Engineering Task Force (IETF) created MGCP to address some of the perceived shortcomings of H.323. See Figure 2-18.

The main purpose of MGCP is to place control of call signaling and processing intelligence in call agents or media gateway controllers. (Call agents and media gateway controllers are synonymous with and similar to the gatekeeper functions in H.323 and are also called soft switches.) A new version of MGCP, released in August 2000, is called Megaco or H.248.

Although Megaco was created for the same purpose, Voice-over-IP, it differs from MGCP because it supports a broader range of networks and devices such as ATM, Remote Access Servers, Multi-Protocol Label Switching routers (MPLS),

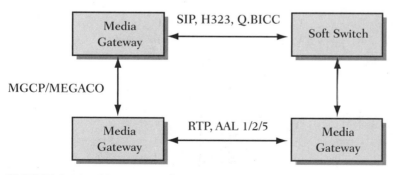

FIGURE 2-18   Megaco interface.

Digital Subscriber Line Access Multiplexers (DSLAMs), and more.

Because Megaco is very new, interoperability testing is ongoing. It appears to answer many of the deficiencies of H.323 and will become very important as we move toward Voice-over-IP networks. It is somewhat unclear at this time if Megaco will replace MGCP or just supplement it. Megaco is more suited for media applications than MGCP, but MGCP may be a better choice for nonmedia-centric applications, such as MPLS-based session control.

SESSION INITIATION PROTOCOL (SIP)

The Session Initiation Protocol (SIP) is an application-layer control protocol that can establish, modify, and terminate multimedia sessions or calls. Like MGCP, SIP is text-based. SIP came out of the Internet Engineering Task Force (IETF) in 1998 as an RFC. It has rapidly gained widespread support, including Microsoft's announcement that SIP will be supported in the next generation Windows XP product.

SIP uses a "request-response" model like that used in Hypertext Transfer Protocol (HTTP). There is one major difference between MGCP and SIP—a call agent is not necessary to mediate between clients. An SIP interface is shown in Figure 2-19.

The usefulness of SIP for multimedia is almost limitless. Sessions can be unicast or multicast and include multimedia

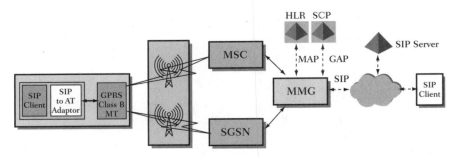

FIGURE 2-19   SIP interface.

conferences, distance learning, VoIP, or similar applications. Some examples of multicast protocols include email, news groups, Web pages, and the like. SIP also supports the ISDN and Intelligent Network telephony subscriber services for personal mobility, which is important for the Wireless Internet.

SIP is reliable, scalable, and can be used with other protocols. Development is fast because it is very similar to HTTP, thus making the addition of feature-rich applications very quick to implement. Initially H.323 and MGCP may be the protocols of choice for tomorrows' media gateways and soft switches but SIP and Megaco will be strong contenders as they mature.

No matter which protocol or protocols become the favorite, soft switches will be flexible enough to adapt. This adaptability makes network service providers very happy. Unlike old legacy switches, this new breed of switches will be quick to accept added features or changed services without waiting months for a manufacturer to modify the switch design.

THE WIRELESS INTERNET MARKET

Forecasts for the Wireless Internet market span a wide range, however, all are in agreement that while the market is still in its infancy, it is poised for enormous growth over the next several years, due to large numbers of people who will have access to it with their handheld devices. Other platforms besides mobile phones, personal digital assistants (PDAs), and Pocket PCs will be able to access the Wireless Internet, appliances, motor vehicles, and other machines will also play a significant role. For the Wireless Internet to be successful, it is important to educate the market of its possibilities and develop compelling applications that will take advantage of devices that can accompany people and provide "anytime, anywhere" access.

INTERNET AND BROADBAND

The United States leads the world with over 167 million Internet users of which more than one third have broadband connectivity, according to a study by Arbitron. This study concluded that 64 percent of Internet users who have broadband access were connected through their workplace, and the balance have home access. The Internet has transitioned from text-based email and file transfer in the mid-1990s to digital audio broadcasting and animated advertising in early 2000. The demand for high bandwidth applications such as animation, video streaming, and high-speed connections to corporate networks has led to a multi-megabit bandwidth race. Digital subscriber line (DSL) and cable modem technologies are main contenders for this high bandwidth race, followed by emerging wireless networks.

In the autumn of 2000, the U.S. Department of Commerce found that 51 percent of all American homes had a computer, and that nearly two-thirds of American Internet users had bought something online. These percentages are similar to those of other developed nations, as depicted in Table 3-1.

TABLE 3-1   Internet Access by Country

PERCENTAGE OF PEOPLE WITH INTERNET ACCESS AT HOME OR WORK (AGE 16+)	ACCESS AT HOME	ACCESS AT WORK
Country	Percent	Percent
Australia	50	30
Austria	42	27
Belgium	39	23
Denmark	58	38
Finland	49	37
France	22	17
Germany	35	22
Hong Kong	58	23
Ireland	46	25
Italy	34	14
Netherlands	56	28
New Zealand	51	31
Norway	53	38
Singapore	56	21
Spain	20	11
South Korea	57	17
Sweden	61	41
Switzerland	43	31
Taiwan	50	19
UK	46	26

Source: Nielsen/NetRatings

By 2001, there were over 2.3 million DSL customers in the United States. Estimates show the market for DSL customers will reach 7.74 million residential customers and 1.83 million business lines, for a total of 9.57 million DSL lines deployed by 2003.

High-speed connectivity is a commodity. As service providers adapt their networks and strategies to meet the demand for high-bandwidth services and applications at reduced cost, some are attempting to capture high-bandwidth customers by offering access to key information service providers (e.g., a specific Internet service or digital video provider).

The customers' key interest in high-bandwidth service includes speeding up large file transfers, viewing high-resolution images and enabling multimedia applications such as streaming audio and video. The attempt of service providers to restrict access to certain information content providers is likely to divert attention from developing cost-effective high-speed networks and create opportunities for other providers to offer access to other information content providers.

Early adopters of broadband services are affluent. Of broadband users, 21 percent of these households have annual incomes over $100,000. Broadband users are also twice as likely to be active online purchasers when compared to low-speed users.* Additionally, most of these users also have cellular phones, PDAs, and other handheld devices such as digital cameras and music players.

In a recent survey of online consumers, 80 percent stated that they would pay approximately $25 per month for broadband access alone; 26 percent of those said that they would pay $50 or $60 for a package of broadband-enabled applications (including premium quality downloadable music or video files) in addition to the cost of broadband access.† Tables 3.2 and 3.3 show the number of users and market penetration of wireless access worldwide.

As cellular technology connects phones, PDAs, and other devices across networks, wide area networks (WAN), local area networks (LAN) and the personal area network (PAN), we see the emergence of IEEE's 802.11B as the de facto wireless

* Jupiter Communications Consumer Survey.
† Jupiter Communications 1999.

WAN standard. As major corporations including Cisco, Compaq, Dell, and others are designing their products for faster cable-free network setup in offices and corporation campuses, we believe that the home market will also adapt, according to a study conducted by the Aberdeen Group, the home network market, including both wireless and wired initiatives will hit $13 billion by 2005.

TABLE 3-2   Top International Wireless Markets 2000

NUMBER	MARKET	SUBSCRIBERS	PENETRATION (%)
1	USA	93,650,000	34
2	Japan	57,950,000	43.36
3	China	46,500,000	3.7
4	Italy	31,118,321	52.2
5	South Korea	27,500,000	53.3
6	United Kingdom	25,517,000	42.9
7	Germany	25,000,000	30.2
8	France	21,082,000	35.5
9	Spain	16,370,150	41
10	Brazil	14,438,963	8.3
11	Taiwan	11,452,541	51.6
12	Turkey	9,234,976	14
13	Mexico	8,694,500	8.6
14	Australia	7,824,560	40.8
15	Netherlands	7,139,000	44.9
16	Canada	7,000,000	22.4
17	Sweden	5,353,000	60
18	South Africa	5,300,000	12.2
19	Portugal	4,804,671	47.8
20	Argentina	4,683,522	12.7

Source: refreq.com

TABLE 3-3   Top Cellular Carriers USA 2000

NUMBER	COMPANY	SUBSCRIBERS	NUMBER OF MARKETS	TECHNOLOGY
1	AT&T Wireless	11 million	105	TDMA
2	SBC Wireless	10.3 million	167	TDMA
3	Vodafone Air touch	9 million	150	TDMA/CDMA
4	Bell Atlantic Mobile	8 million	75	CDMA
5	GTE Wireless	5 million	141	CDMA
6	Alltel	5 million	265	CDMA
7	BellSouth Wireless	4.9 million	93	TDMA/CDMA
8	United States Cellular	2.6 million	139	TDMA
9	Western Wireless	840,000	96	CDMA
10	Century Tel	708,000	44	TDMA
11	Dobson Cellular Systems	661,000	67	TDMA/CDMA
12	Centennial Communications	526,000	31	TDMA/CDMA
13	Price Communications	450,000	16	TDMA
14	Rural Cellular Corp.	260,000	20	TDMA
15	Trito Cellular Partners	210,000	20	TDMA
16	Wireless One Network	180,000	6	TDMA
17	Cellcom Cellular	165,000	8	TDMA
18	Midwest Wireless	160,000	12	TDMA
19	Plateau Wireless	70,000	5	TDMA
20	Bluegrass Cellular	66,000	3	TDMA

WIRELESS SUBSCRIBERS AND INTERNET GROWTH

This new market is powered by fast-growing demand for mobile and Internet services and complementary technologies that allow people and information to be increasingly interconnected. The penetration of wireless service subscription continues to increase dramatically around the globe. It is almost impossible to be in a place where no one uses wireless phones. Continued growth in the United States, Europe, and Japan is strengthened by emerging wireless device industries in China, India, and Latin America. Riding on this wave of growth, the number of wireless subscribers by mid-2001 exceeded 119 million subscribers in the United States alone, according to the CTIA, and the number of U.S.

Internet users topped 167 million based on Nielsen NetRatings. Furthermore, over 60 percent of U.S. households have online Web access. The Strategies Group predicts that wireless data penetration will reach 60 percent in 2007, from just 2 percent in 2001. This massive growth is attributed to the aggressive rollout of high-speed services and applications and consumer acceptance. Furthermore, over the next several years, the majority of devices that tap the Internet will not be home or office PCs, but rather wireless devices. This trend is accelerated in Japan and Europe, where the Wireless Internet is transitioning from text-based short messaging (SMS) to digital audio and video broadcasting. The demand for high-bandwidth wireless applications and connections to corporate networks from the field is fostering the development of and demand for Wireless Internet networks. Figures 3-1, 3-2, 3-3, and Table 3-4 break down United States and global Internet and wireless use.

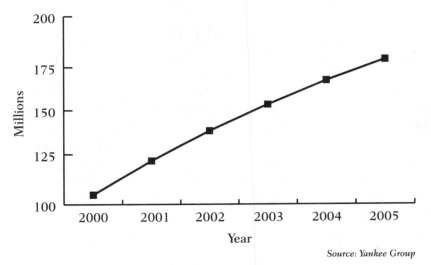

Source: Yankee Group

FIGURE 3-1   Forecast total U.S. wireless phone subscribers.

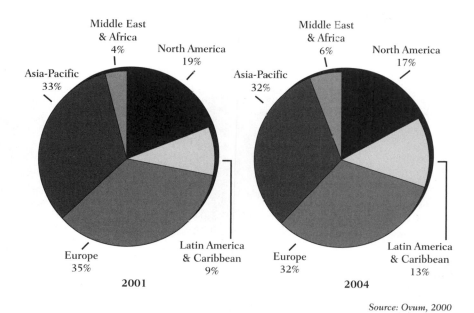

Source: Ovum, 2000

FIGURE 3-2   Percentage of global mobile subscriptions by region.

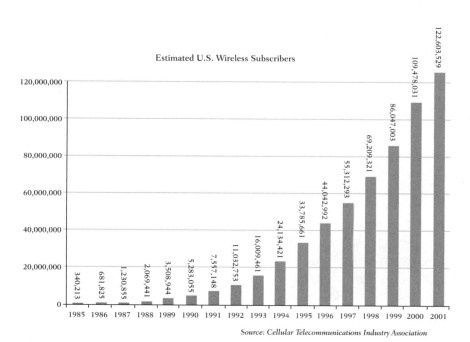

Source: Cellular Telecommunications Industry Association

FIGURE 3-3   CTIA's semiannual United States wireless industry survey results—December 1985–September 2001 (a historical perspective).

TABLE 3-4   United States Online Users

DATE	USERS	PERCENT OF POP.	SOURCE
December 2000	164.4 million	59.86	NielsenNetRatings
November 2000	153.84 million	53.8359.66	NielsenNetRatings
October 2000	149.6 million	54.29	***NielsenNetRatings
September 2000	148.03 million	53.72	***NielsenNetRatings
August 2000	146.9 million	53.31	***NielsenNetRatings
July 2000	143.96 million	52.24	***NielsenNetRatings
June 2000	134.2 million	48.7	***NielsenNetRatings
February 2000	123.6 million	45.33	*** NielsenNetRatings
January 2000	122.8 million	45.04	*** NielsenNetRatings
July 1999	106.3 million	39.37	*** NielsenNetRatings
May 1999	101 million	37.4	*** NielsenNetRatings
April 1999	95.8 million	35.4	*** NielsenNetRatings
April 1999	92 million	34	* CommerceNet/Nielsen
March 1999	83 million	30.7	* IntelliQuest
January 1999	79.4 million	29.3	* IntelliQuest
October 1998	73 million	27.8	* IntelliQuest
August 1998	79 million	29	* CommerceNet/Nielsen
February 1998	62 million	23.0	* IntelliQuest
November 1997	56 million	21.0	* IntelliQuest
June 1997	51 million	19.17	* IntelliQuest
April 1997	40 – 45 million	16.16	* FIND/SVP
1995	18 million	6.7	* CommerceNet/Nielsen

  * Figures quoted are for adult population only (age 16 and over). They do not include number of children online.

  ** Figures are for Internet users age 12 and older.

*** The Nielsen/NetRatings Internet universe is defined as all members (2 years of age or older) of United States households that currently have access to the Internet.

The growth of the Wireless Internet is directly linked to the success of the wired, HTML-based Internet. The Yankee Group, a major research firm, estimates that by 2005, approximately 56 million people in the United States, or almost 20 percent of the population, will regularly tap into the Wireless Internet over a voice-enabled device; a major investment banking firm takes an even more optimistic outlook, projecting 115 million subscribers in the same period. The Wireless Internet will exploit the gold mine of content available in digital format from Internet servers designed originally for wired desktops. Many of the largest players, such as AOL, Yahoo!, and Microsoft are charging forward into wireless as a key growth initiative. Carriers such as AT&T, Nextel, Sprint PCS, Verizon, OmniSky, Metricom, Vodafone, and many others offer Wireless Internet access through cellular phones, PDAs, RIM pagers, and various other devices. Figure 3-4 shows projected Wireless Internet growth in the United States and Figure 3-5 shows projected growth worldwide.

COMPUTER SALES

In 2000, according the IDC, approximately 101.4 million desktop PCs were shipped, generating $141 billion in revenues. Many new consumers were attracted by low prices, higher capabilities, and stylish colorful designs. During the same period, a total of 26 million notebook computers, valued at $57.7 billion, were shipped worldwide, based on IDC research. Additionally, most PC vendors support at least one wireless protocol and many offer services to facilitate the transition to a wireless computing environment. It is projected by many lead-

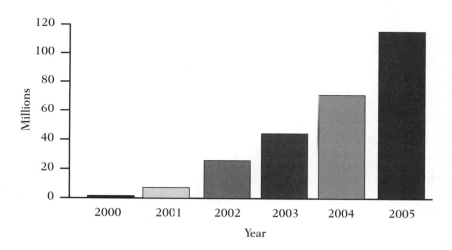

Source: Investment Banking Reports

FIGURE 3-4 Estimated number of United States Wireless Internet users, using SMS, WAP, email, or browsing at least once per month.

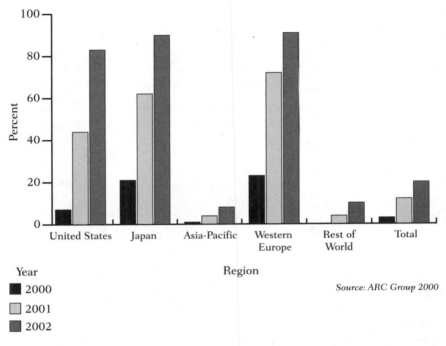

FIGURE 3-5   Penetration of mobile data users by region.

ing market research firms that within the next two to three years, the desktop PC market will decline in favor of wireless-enabled portable PCs, handheld Pocket PCs, and other devices.

Wireless phones, PDAs, pagers, and modems are expected to surpass PCs as the most popular Internet access devices. Shipments of Wireless Internet devices will sustain double- and triple-digit growth over the next few years. Mobile electronics sales may rise to $10.5 million in 2001, as consumers continue to enjoy the ease with which they can access information and entertainment and stay in touch with friends and family. Wireless phones maintain the largest share of that category, with sales forecast at $3.7 billion in 2001, a 16 percent increase over 2000, according to the Consumer Electronics Association (CEA). While the PDA market in the United States continue to be strong, the Yankee Group projects 13 million to be sold in 2001, growing to over 26 million in 2003. Approximately 11 percent of PDAs had Internet access, while

Jupiter projects that the number of U.S. PDA users interacting with Web data and content will approach 14 million by 2005.

Major consensus holds that within the next three to five years, the majority of devices that tap into the Internet for data will not be wired personal computers, but rather a mix of hand-held devices—including cellular phones, PDAs, pagers, and Pocket PCs. Research firm IDC projects that the worldwide market for these devices will grow to over 67 million units sold and $18.1 billion in revenues by 2005. Many of these devices shall also offer to the consumer new and integrated features and functions such as cameras and music players, which will make it more convenient for consumers to carry only a single device. Many of these devices will take on new form factors—designs which look less like a phone. These new devices will be designed to make viewing and entering information easier.

This same period will also witness the emergence of human-to-machine and (though still embryonic), machine-to-machine communications, provided by embedded wireless communication links for data exchange. This will improve work flow for the business user and create new opportunities for companies. This trend accelerated in Europe and Japan. Examples of human-to-machine communications via wireless networks already exist, allowing people to access content and applications from network servers. One example of machine-to-machine communications is vending machines. A wireless device embedded in the machine communicates with a central computer that keeps track of how many soft drinks or other items are left; it lets suppliers know when the vending machine needs to be restocked. The device may also notify the central computer when the vending machine is in need of repairs. These devices can be attached to home appliances such as refrigerators, air conditioners, and security systems to control lights, activate alarms, and provide climate control. Other embedded devices will monitor environmental factors like carbon monoxide levels. In the vehicular environment, these devices will provide navigation aids and also work as security and theft-prevention devices. Referred to commonly as telematics, which is the blending of computing and wireless telecommunication systems, which creates useful applications for automobiles and trucks. Telematics systems often use global positions systems (GPS) or cell-based technologies to facilitate location-based services such as roadside assistance offered by companies such as GM's OnStar. For telematics to become successful and widely accepted, car manufacturers have to agree on a standard for hardware and operating system, otherwise companies have to build specific versions of their applications for each auto manufacturer, resulting in a segmented industry.

In the United States alone, there are over 210 million existing cars and sales of new cars are about 17 million per year. Worldwide annual sales of new cars are over 50 million. According to the Yankee Group, 50 percent of new cars and 90 percent of highend vehicles will have telematics to keep users connected while in the car in 2006, which equates to a market over $25 billion. Furthermore, on the enterprise side, there are over 40 million fleet vehicles in the United States alone, which includes trucking, delivery, and service vehicles. We believe that in the not-to-distant future, connectivity with the Internet for many consumers will be via wireless device and not home PC.

The U.S. mobile phone market is expanding with an additional 11 percent of households expected to purchase a wireless handheld device in 2002, according to a survey conducted by Takar Nelson Sofres Intersearch. We find that over 40 percent of Nextel's subscriber base has data-capable handsets and more than half of Sprint PCS subscribers have data phones. These run text-based microbrowsers that can only surf limited numbers of Web sites, but that number is growing. Using *push technology*, it will be possible for sites to alert users of downloadable content, based on customized settings. These early trailblazers will fundamentally change Internet usage patterns from one or two long sessions a day on an office or home PC to dozens of low-intensity sessions a day for specific Web clips or data points. (Figure 3-6 shows projected U.S. data-only service growth, while Table 3-5 shows worldwide handheld shipments.)

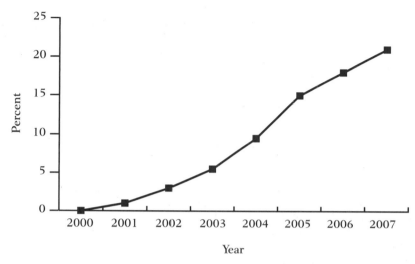

FIGURE 3-6 Projection of U.S. data-only services penetration rate.

VOICE INTERACTION: LINKING THE WEB BY VOICE

Voice portal services began in 1999 with a handful of vendors, including General Magic and Wildfire, forging the way. Since then, several dozen vendors have appeared. Most of these services were initially offered free of charge. Wireless devices are increasingly using speech technology as an alternative user interface to access applications.

Excite, Yahoo!, Tellme, and many other Web sites allow people to communicate using both voice and text. Users can check email; receive voice messages; and access news, weather, stock quotes, and sports results from their phones. Additionally, many handheld devices are incorporating voice interaction technologies to aid in navigation and operation of the device. Voice is the simplest human-to-machine interface, and as such can become a standard way to navigate and enter data on wireless devices, whereas a visual display will probably remain the preferred way to receive and view it. Acceptance of speech as a user interface will vary by region because of cultural and social factors. (The safety issue of operating a handheld device while driving a vehicle should cast a favorable light on use of speech

TABLE 3-5   Worldwide Handheld Shipments (Figures in Thousands)

	1998	1999	2000	2001E	2002E	2003E	2004E
Total Mobile Devices	**168,413**	**295,348**	**416,249**	**439,314**	**531,316**	**573,344**	**652,124**
Mobile Phone Handsets	163,424	289,543	405,000	423,582	506,099	535,236	596,156
Handsets Without Browsers		237,649	328,255	232,970	177,135	107,047	59,616
Mobile Handsets with Browsers		51,894	76,745	190,612	328,965	428,189	536,540
Mobile Total Handsets (%)		17.9	18.9	45.0	65.0	80.0	90.0
SmartPhones	655	555	480	1,418	3,901	9,556	20,546
Handheld Data/PDA	4,334	5,250	10,769	14,314	21,316	28,552	35,421
Total Smart Handheld Devices	**4,989**	**5,805**	**11,249**	**15,732**	**25,216**	**38,108**	**55,967**
Integrated Wireless (%)	*0*	*2.7*	*3.6*	*5.3*	*8.5*	*12.3*	*16.4*
Integrated Units (000)	*0*	*142*	*388*	*759*	*1,812*	*3,512*	*5,809*
Add-on Wireless (%)	*0*	*6.3*	*7.9*	*11.5*	*15.8*	*22.3*	*28.2*
Add-on Units (000)	*—*	*331*	*851*	*1,646*	*3,368*	*6,367*	*9,989*

Source: IDC, GartnerGroup, Company Reports, and U.S. Bancorp Piper Jaffray Estimates.

recognition technologies.) Personalization will be key for customer adoption of voice processing technology. Just as users personalize their information on the Web, customizing their voice portal will reduce lengthy set-up menu processes and permit users to quickly move to the desired content. Better delivery and performance and more dynamic content will contribute to greater acceptance. According to Giga, voice recognition services will grow at 70 percent annually, and revenue from voice portal applications will grow to $45 billion by 2005.

MARKET CASE STUDIES

EUROPEAN AND JAPANESE ENVY

Why does it seem that most of the rest of the world has a better communications system? We often hear that Europe and Japan are ahead in terms of wireless technology; surprised Americans have a hard time understanding why they don't have the latest in devices and systems. Although there are many factors that have contributed to the apparent head start of Europe and Japan, one of the most important is geographic size and population density. In short, they simply have less space to cover and often more people in that smaller space.

Cellular gets its name from the system of cell use and reuse that essentially divides up the area to be covered into *cells*. Each cell can only handle a certain amount of traffic before it, too, needs to be divided into smaller cells to increase capacity. The first analog cellular systems employed frequencies that covered a relatively large area while handling a modest amount of traffic. As the popularity of cell phones increased, more and more analog cell sites were needed. Digital 2G systems were designed to handle this increasing traffic by utilizing more efficient digital systems that required less power and could fit more calls into the same amount of bandwidth. These digital 2G systems were not only more expensive that 1G analog systems but required more cell sites. This greatly increased the total build-out costs of service provider network infrastructure.

Because many European countries could fit into the United States several times over, and Japan could fit into Europe several times over, European and Japanese operators could roll out a 2G network with more complete coverage for less money than could an American operator. We see this same effect in the coming 3G systems—is anyone really surprised that Japan will have it first?

Americans shouldn't feel too badly: the United States is the home of the PC-based Internet, whereas Europe and Japan want bandwidth for the mobile phones. By mid-2001 over 65 percent of European Union citizens had a mobile phone, more than twice the percentage who had home Internet access. In some countries, Finland and Sweden for example, mobile penetration levels have reached 70 percent, and there are some predictions that some countries may reach 100 percent penetration (one for every man, woman, child, and dog—yes, even pets can wear communication collars that track their location. Imagine being able to call your pet via the built-in attached speaker). Mobile communications is a global $200 billion industry with a growth rate of 12.5 percent a year. European wireless operators spent an astronomical $130 billion in 2000 on licenses (a staggering amount to spend on "air") to offer 3G services, and will have to spend as much to build out their networks. As global culture continues to merge, disparities will eventually equalize as each nation learns from other's advances.

EUROPEAN EXPERIENCE

Most countries in Europe, specifically the Scandinavian countries and Finland, are focused on communications and mobility. Wireless telephony has been part of European life for most of the past two decades. In these countries, more than 60 percent of the populations use mobile phones. In many cases the use of mobile phones is so popular that it has dramatic impact on the growth of conventional landline phones. Consequently, mobile phones are replacing conventional phones in certain households and lifestyles. As such, it is predicted that by 2004, one third of all Europeans—over 200 million people—

will regularly use Internet services on their mobile phones, according to research company Forrester. As this Internet fever takes hold in Europe, more Europeans are racing to route e-commerce through mobile phones. It's estimated that within the next two years more Europeans will be surfing through a Web phone than a PC. The Wireless Internet is poised to become the most important channel for online banking and financial services in Europe.

Across Europe, new wireless data services and applications are being launched. As operators make the transition from circuit-switched to higher speed data services, they and the market will provide applications that make use of the resources available to further drive usage of the network. They will also have to make a move away from time-based billing and examine the alternative revenue streams that are available to applications providers. In Western Europe, 45 out of 57 mobile network operators in 18 countries, representing approximately 90 percent of all the mobile phone users in Europe, have already created online portals.

FINLAND: THE LAND OF THE MIDNIGHT SUN, REINDEER, AND MOBILE PHONES

In Finland, the number of the households having a conventional telephone has decreased during the past ten years from 94 percent to 78 percent. Simultaneously the amount of households having a mobile phone has shown a very rapid growth from less than 5 percent to 73 percent. By 1998, more Finnish households had a mobile phone than a conventional telephone. Approximately 88 percent of households in Finland have at least one mobile phone, and more than 20 percent of these households have more than one mobile phone, which is up from 65 percent in 1999. More than 20 percent of households in Finland have only a mobile phone (one or more), and that number may grow in a few years. (*Source:* Statistics Finland; www.cellular.co.za/news.)

Finns in particular have a strong affinity to their Nokia mobile phones. In 2000, over 1.4 million new phones were purchased (in a country with just over 5 million people). These phones are used to send messages, especially by teens. In 2000, over 1 billion SMS messages were sent. Furthermore, wireless phones are also being used in a variety of mobile transactions. For example, to purchase beverages from GSM-enabled vending machines, the user dials the number indicated on the machine, which results in the release of the soft drink from the machine. The cost of the beverage appears on a monthly bill, together with the charge for the phone call.

MADE IN JAPAN: THE LAND OF THE RISING WIRELESS INTERNET

The Japanese market has given the world a glimpse of what the Wireless Internet might look like, and it is a prime force in the direction and momentum of the Wireless Internet market. NTT DoCoMo is Japan's leading mobile phone operator and largest ISP and the world's leader in Wireless Internet access. Since starting its data service (called i-Mode) in February 1999, NTT DoCoMo has seen its subscriber base grow to top 26 million as of August 2001; it averages 40,000 to 50,000 new subscribers per day. This spectacular growth is driving NTT to implement 3G technologies that will provide for greater capacity and allow creation of new applications. Only about 15 percent of Japanese consumers and business people access the Internet via PCs. Thus to many of NTT's i-Mode customers, the i-Mode is synonymous with the Internet. According to NTT DoCoMo, the "i" stands for interactive, Internet, and independence.

TABLE 3-6 Forecasted Subscriber Growth in Japan

	1999 (IN MILLIONS)	ESTIMATED 2005 (IN MILLIONS)
3G	0	25
1S-95	1	5
PDC/PHS	48	50
ANALOG	6	<1
TOTAL	55	80

Source: Lehman Brothers

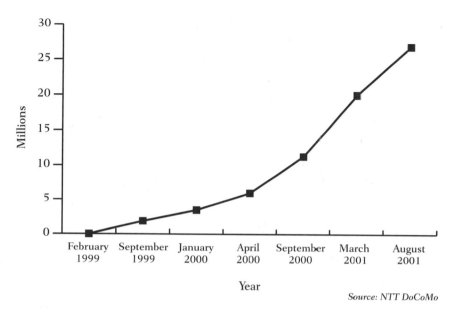

Source: NTT DoCoMo

FIGURE 3-7   NTT DoCoMo's i-Mode subscriber growth (February 1999–
August 2001).

In Japan, businesses from airlines to television stations to
banks all provide their services via i-Mode. DoCoMo earns a 9
percent fee from content providers that charge for their information. DoCoMo has four revenue models: monthly subscription

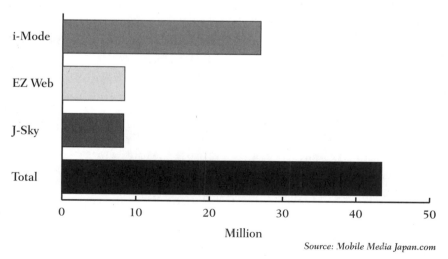

Source: Mobile Media Japan.com

FIGURE 3-8   Japanese mobile net subscribers.

fees, per-packet data transmission fees, commissions on billing, and revenues for traditional voice services. In 2000, DoCoMo reported that 70 percent of i-Mode traffic came from users in their teens to early twenties, with about 60 percent of i-Mode traffic going to official sites that specialize in entertainment. According to NTT DoCoMo, the average total bill for i-Mode data transmission is about $13.00 (U.S.) per month. Equally impressive, the average i-Mode user generates an additional 36 percent increase in revenue over voice-only subscribers. Much of this increase can be attributed to direct access, use of data packets, and increasingly higher voice minutes of use.

The increase in voice usage is interesting in that we believe it represents a hidden upside to most United States business models. Even more impressive is i-Mode's transmission speed of only 9.6 Kbps. Even at this "slow speed," some of the best selling applications are (surprisingly) cartoon-character screen savers that download each day for $1. The i-Mode service has been so successful that at times DoCoMo has curtailed its advertising in efforts to slow down subscriber growth while improving network capacity. When DoCoMo launched i-Mode, it had 67 Web site providers: By the end of the first year, there were 721 information providers responsible for 1,280 sites on DoCoMo's main i-Mode menu, and third party developers had created another 31,085 i-Mode sites. Additionally, it was announced in February 2001 that that Google had developed a new technology that gives i-Mode users in both English and Japanese access to the more than 1.3 billion Web pages Google has indexed to date. Google's technology converts a request for a standard HTML Web page to be viewable on an i-Mode handheld device. i-Mode's success is enhanced by the huge number of content sites available to the subscriber.

To better understand the reason for i-Mode's popularity, and the rapid and overwhelming adoption of the Wireless Internet by the Japanese people, we must look at Japan itself. In Japan, space is at a premium—homes and offices are small and there is very little extra room to accommodate PCs, monitors, and printers. Furthermore, Japanese society is traditionally an early adopter of technology in general and it is a commuter culture.

FIGURE 3-9   i-Mode phone. Courtesy of Mitsubishi.

Today, only about 15 percent of Japanese consumers and business people reach the Internet using PCs. The remaining 85 percent are willing to accept the limitations of smaller display screens and keyboards on wireless handheld devices. Furthermore, the price of PC Internet access via landline phone is higher in Japan when compared to the United States or Western Europe. The average costs are $20 per month plus $2 per hour of use. The installation price of a home phone line is approximately $700, as compared to a cellular connection for $28; and i-Mode users pay only for the number of packets used.

Because the Japanese are traditionally early adopters of new technologies, they have been very quick to adopt new i-Mode products and services. For example, Japanese consumers have purchased dog collars that transmit their animal's location to their wireless device, PC, or fax machine. Entertainment-related sites where you can download images, ringing tones, play interactive games, read your horoscope, find dating services, weather, and news are most popular. Because the majority of

Japanese students and employees commute (usually by train or bus) an average of 30 to 40 minutes per day each way, they have ample opportunity for mobile communications and entertainment. Pocket-size devices are really well-suited to these commuter environments where, more often than not, there is no room to even open a newspaper. Even a small notebook computer is too big to carry on a bus or train, whereas the i-Mode device is the perfect size to be held in one hand. Table 3-7 shows the popularity of i-Mode content by category. Figure 3-10 shows preferred mode of access to the Internet in Japan.

TABLE 3-7   Popularity of i-Mode Content

CONTENT TYPE	PERCENTAGE	EXAMPLES
Stored in Database	13.6	Dictionary search, remote mail, restaurant guides, recipes, telephone directories, city information
Transactions	20.7	Airline ticket and hotel reservations, credit card bill inquiry, stock searching and trading, balance checking, money transfer, bill payment
News and Information	13.2	Business news, television listings, sports news and weather forecasts
Entertainment	52.5	Network games, downloading avatars and game characters, fortune telling, karaoke, downloading ring tone melody and FM on-air information.

Source: NTT DoCoMo based on total number of hits.

Because of the unique characteristics of Japanese society, it remains to be seen how quickly other societies and cultures embrace the Wireless Internet experience. Acceptance will depend on packaging and pricing, and quality and quantity of compelling content and services. However, as the Japanese experience clearly demonstrates, acceptance of the Wireless Internet is high when things are done right.

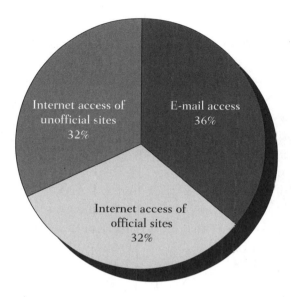

FIGURE 3-10   Mode of access by type of service.

THE NEW GENERATION OF CUSTOMERS

TEENAGERS DRIVE WIRELESS INTERNET GROWTH

Much of the demand for Wireless Internet usage comes from teens and young adults. Wireless Internet growth will be driven by people growing up with the Internet and wireless communications as an integral part of their lives. This is especially true among the younger segments of society, where the adoption rate of new technology is rapid. That generation of children growing up with access to the Internet throughout the United States, Western Europe, and Japan is incredibly adept at using technology, whether through school work, video games, chat rooms, buddy lists, or short messaging. As this Internet-literate generation joins the workforce, their spending will increase, as will their acceptance for newer, more enabling applications and devices. Jupiter found that teenagers represent 12 percent of the European online population, and in June 2001 Europeans ages 12 to 17 spent nearly 8 hours online. As this trend continues, Web sites must be designed for the specific characteristics of their target audience: boys seek novelty and entertainment, whereas girls enjoy fulfilling goals and feeling part of a community.

A research study from the Pew Internet and American Life Project has found that 73 percent of U.S. teenagers aged between 12 and 17 (or 17 million people), use the Internet. Furthermore, three-quarters of online teens say they would miss the Internet if they could no longer use it, whereas almost half say being online has improved their relationship with friends. Nearly a third say it has helped them to make new friends. The top five online activities for teens are email, surfing for fun, visiting entertainment sites, using instant messaging (IM), and researching hobbies. Only 31 percent have made purchases online. About 13 million teens, or 74 percent of all online teens, use instant messaging (IM). Only 44 percent of online adults use IM. Nineteen percent of teen IM users say IM is now the main way that they communicate with their friends.

When it comes to wireless phones, teens just can't keep their hands off them and stop talking. It is predicted that by 2004, more than half of U.S. youth, over 43 million, will own a wireless phone and three out of four will use one. Teens want the flashiest, most featured models— a stripped down phone is "just not cool" nor is a grey or black model that looks just like Dad's. Teenagers are very fashion conscious, and major manufacturers such as Nokia, Motorola, and Ericsson are all focusing on new designs that appeal to the young generation. Both carriers and application developers are focused on specially targeted content, including sites providing shopping, news, games, entertainment, education, and youth-oriented content.

WIRELESS BUSINESS ENTERPRISE

Over the next five years, corporate users will join consumers in the wireless service user base. This trend has already started as companies use wireless technologies to connect corporate information technology (IT) systems with customers, employees, suppliers, and partners. Companies realize the benefits of providing wireless access to email, instant messaging, portals,

FIGURE 3-11   A colorful cell phone. Courtesy of Ericsson.

and corporate systems. These benefits include reduced administrative overhead, increased efficiency, and a more rapid distribution of information throughout an organization.

For mobile professionals, a wireless device allows access to all sales literature, provides answers to questions about unfamiliar products or services, and permits check-in with the home office for timely reports, expenses, and inventory status, among other things. Enterprises are using wireless devices with an Internet browser to provide remote access to corporate data—up-to-date access to recent sales, current orders, proposals in progress, and accounts receivables. The key to ensure the acceptance and usability of these devices is to design them be customized for the precise information needed by the user; thus, only relevant information is displayed rather than the wealth of information usually displayed on management information systems.

IN A NUT SHELL...

The potential market for the Wireless Internet, by all accounts, will be a substantial and sustainable one as people embrace what may be gained by anywhere, anytime connectivity to the varied content found on today's Web.

The Wireless Internet will most certainly provide society with new experiences and freedom, and unprecedented access to information. The Wireless Internet frees consumers from location and time constraints, making it possible to shop for merchandise or initiate transactions from virtually anywhere, day or night, without sitting in front of a PC.

Because the number of wireless devices exceeds the number of home and office PCs, and this gap will widen in the coming years, the impact of wireless users will be the major driver for future content. Traditional Web sites will adapt to accommodate the various screens of many handheld devices, each of which will be designed for various applications. The Wireless Internet untethers an endless stream of information, new solutions, and opportunities for human communications by offering wireless connectivity to the vast knowledge and resources of a networked world.

WIRELESS INTERNET APPLICATIONS, SERVICES, AND ACCESS-ENABLED SOLUTIONS

Wireless Internet applications are software programs that require wireless communication technology that can take advantage of the mobility and high-speed data transmission offered by advanced data services and networks. Many of the communications applications and services that were available for mobile communications in the 1990s were limited by slow-speed (less than 10 Kbps) data transmission. Using 2G mobile systems, it was not possible to offer streaming video, rapid image file transfer, or high-speed data file transfer services. New high-speed networks will enable applications that process images, color, and moving video to provide users with a far richer experience than possible using voice alone.

Much of the demand for wireless data access has come from a combination of the availability of Internet information applications and low cost mobile communication. The Internet's standardized global collection of interconnected computer networks has allowed access to information sources that provide significant benefits to those companies and individuals looking for specific knowledge. The Internet has created a culture-changing awareness of many new information services.

In the late 1990s, new, low-cost, high-speed connections to the Internet became available. The resulting rapid market growth of Digital Subscriber Line (DSL) and cable modem technology has stimulated the development of new applications that are only possible via broadband high-speed connections. In the early twenty-first century, consumers are becoming aware of these new broadband multimedia applications and the transition back to low-speed text-based services is difficult. In the near future, as high-speed wireless networks are deployed, cellular phones and PDAs will be converted into portable stereos and video players. These new combined devices eliminate the need to carry various gadgets, offering consumers convenience.

Already many consumers are aware of the benefits of wireless mobile service and broadband applications. Potential Wireless Internet customers may only need to be made aware that these services can be delivered via high-speed wireless data communication services to convert them from the traditional wired (e.g., Internet access) to new wireless services.

Of key importance for 3G technologies are those broadband applications that provide the mobility, low cost installation, and rapid deployment that competing broadband technologies cannot provide.

ACCESS VERSUS APPLICATIONS

Remote access and more specific wireless applications will be key drivers for wireless industry growth in the next decade. Accompanying this phenomenal growth are substantial opportunities for those mobile operators, equipment manufacturers, and developers that can answer end users' demand for customized value-added services, applications, and content. With the convergence of communications and computing evolving into the next generation, 2.5G and 3G wireless broadband companies, service providers, devices manufacturers, and content developers are focused on identifying and developing the "Killer App."

It's important to clarify the difference between wireless Internet access or transport and wireless applications. Access or transport is simply a wireless connection to the Internet that allows users to access the same content and applications they would from a fixed-wire connection. Wireless access adds value to users of those laptop PCs that are capable of displaying and running applications that have been designed for PCs. This type of wireless access allows many new methods of conducting business in places that do not have wired connections available such as remote job sites, airports, and in cars.

Whereas mobility adds value to the Internet, not everyone would consider a laptop to be the most convenient mobile device to carry with them. Size and weight are important to mobility and therefore smaller devices are more suited to mobile usage although not as capable for running applications designed for the large screens, abundant processing, and large storage enjoyed by the typical personal computer.

Smaller devices such as smart cellular phones and PDAs are much easier to carry and are small enough to fit into a pocket or purse with little difficulty. Small size is great for mobility but not so good when it comes to using and viewing applications made for PCs. For the wireless customer, *gateways* and *middleware* players are being developed to enable content from traditional Web sites to be usable and viewable on wireless devices. *Vertical applications* are being developed that are specific to users' wireless needs. Many of these new information content services and applications are specifically tailored to a fast-paced, mobile lifestyle that provides for accessibility anytime, anywhere.

The Wireless Internet enables applications to fulfill the needs or desires of end-users with a variety of smaller devices. And although these devices have "access" or connectivity to the Internet, it's the applications that make the difference.

Consumers do not care or need to know how underlying communication and Internet technology networks function or be further confused with various standards; they just care that an application does what they want. An enormous amount of marketing research has been conducted to better understand what content is useful and which applications can best display and interact with this content. Progress has been made in wireless application development but much more is yet to be done to truly understand how to provide value to users of wireless devices.

APPLICATION AND SERVICE CATEGORIES

There will be no single "killer application" for Wireless Internet service because there are many high-value user applications. These services includes:

- Personalized communications
- News and information
- Entertainment and lifestyle
- Location-based services
- Access and connectivity services

MOBILITY VALUE

As the Wireless Internet adds new value to staying connected, short response times assure the validity of information. Productivity is no longer confined to a specific location. There are potentially hundreds of applications that will drive the demand for Wireless Internet access, including *multimedia messaging*, which will make it possible to combine conventional text messages with richer content types—photographs, images, voice clips, and eventually video clips. Two of the fastest-growing industries in the world—entertainment and mobile communications—may profit hugely as lifestyles change, with people experiencing more and rapid bursts of free time. Fast access to entertainment is increasingly appealing to all sections of society and many wireless handsets and devices are already used for entertainment. SMS services led the revolution in enjoying entertainment on the move, but we are now on the edge of a new era, as the Wireless Internet begins to offer even more sophisticated services.

Increasing demand for Wireless Internet access comes from the convergence of Internet and cellular telephony. The Internet or World Wide Web (WWW) has had a profound impact on our lives, both on a personal and business level.

ADDING VISION TO VOICE

When the average consumer thinks of wireless technology most have a difficult time imagining anything but voice services. We are conditioned to think of cellular or wireless devices as things you talk into and not look at or read.

Even the youngest users quickly understand how to use today's cellular phones—dial and talk. It's not uncommon to observe young children grabbing and chatting on a parents' cell phone, even if it's not really turned on. The point is that they get the concept of wireless voice. Unfortunately most cell phone users (even non-toddlers) do not use the many voice-related features embedded in their phones, much less envision how they might someday use them for nonvoice applications.

Wireless devices will start with basic features and gradually evolve as technology and end user perceptions grow. This is not unlike the evolution of the personal computer. Many of you will remember when computers were very limited, and you have witnessed the evolution from a text-only device with one method of input, through the stage of simple graphics and crude sounds (buzzers!), and finally to the full a multimedia PCs of today.

Applications track the capabilities of devices and networks as well. Early PCs were connected to nothing more than the electrical outlet on the wall and were capable of sharing only via the popular "sneaker network"—put the files on a floppy (remember those 5.25-inch disks that truly were "floppy"?) and walk them to the intended destination.

PC applications moved from simple text menu–driven programs to applications that use color, pointing devices, multimedia images, and concert quality sound—and these are just the tax preparation programs! Modern day computer games are capable of experiences that are closely approach virtual reality.

Wireless Internet devices will follow the same path, albeit in a condensed timeframe. The first nonvoice wireless applications will be those that use simple text—displays have become larger with better resolution to accommodate even this simple text. The next stage will be very simple graphics not unlike the simple graphics first experienced on PCs—if you look too closely, you'll see the same jagged edges and low resolution. As network data speeds, device processing power, memory, and displays improve we will see wireless applications that can take advantage of still images, sound files, and finally the 3G vision of fully wireless multimedia.

This gradual evolution is important for the adoption of applications. In part this will allow consumers to experiment and learn how best to utilize this new method of Internet access. It's difficult for most consumers to assess the value of wireless applications that they have not yet experienced. In turn, this makes it hard to determine exactly what applications users will be willing to pay for and even harder to understand *how much* they will pay. Even applications that enable relatively obvious services such as weather alerts or driving directions cannot be fully appreciated out of the context of a real world usage situation.

EMERGING DEVICES

A broad range of applications for the Wireless Internet will warrant a variety, of access devices. The incumbent handset manufacturers—Ericsson, Motorola, and Nokia—have the benefit of years of experience in building consumer terminals, integrated voice service, and have expertise in next-generation wireless air interface standards. Many of today's products already offer WAP microbrowsers and SMS. Competition will be intense among the many manufacturers vying for market share, and this will ultimately benefit consumers through lower prices.

FIGURE 4-1   iPAQ handheld PC. Courtesy of Compaq.

One of the major challenges for the handset vendors is to design phones with simplified text entry and Web navigation. Typing out text messages on cellular handsets is cumbersome, as is navigating through the menus of many phones. Features such as Tegic's T9 predictive text input are useful but require some practice to become familiar enough to be useful.

A new range of products is emerging that will compete with the traditional handset for a share of the Wireless Internet terminal market. Palm and Handspring are adding wireless functionality to PDAs, as are new Pocket PC handhelds from Sony, Hewlett-Packard, and Compaq. The competition among these devices will be intense as manufacturers jockey for position in this rapidly growing device market segment. All of these devices will help build the momentum of the enterprise market as they become tools for corporate users to access company data and read and respond to emails.

One challenge to creating wireless applications has been the ability to write programs that could be loaded and run on devices with proprietary operating systems. Devices such as cellular phones have traditionally used an operating system that did not allow for new programs to be added, and most manufacturers do not release details that would allow developers to integrate a new application into an existing device's software. PDAs have been easier to develop applications for, because platforms such as Palm and WinCE were created with software developers in mind. The drawback to PDA software has been the lack of wireless connectivity.

As cellular phones and PDAs absorb each other's features and abilities it's becoming harder to tell the difference between a PDA and a phone. PDAs that are capable of connecting to the Internet and making voice calls now compete with cell phones that are able to run and display applications. One thing that these two types of devices share is common software platforms that enable users to add and use wireless applications.

Some device manufacturers are also designing handhelds that integrate other devices such as digital camera, music player, and others that result in some interesting combinations. These new all-in-one, "Swiss Army Knife"-style mobile devices may not necessarily meet the needs of the power user, but will eliminate the need to carry multiple gadgets. For instance, Samsung and Sprint PCS offer an MP3 phone and in Japan, camera phones have been available for the past couple of years. Leading device designers such as the United Kingdom firm of Seymour Powell are planning future devices that may no longer look like today's cellular phones, but take into account people's usage habits, resulting in products that make viewing, listening, entering information, and interaction with the device much simpler and intuitive. While some of these new hybrid devices may be compelling, we do not believe that they will ever completely replace the basic wireless phone. Additionally, any large scale phone replacement cannot occur until the proper wireless network architecture is in place, which we believe will not occur until 2004.

As device manufacturers offer more functionality in their products: color screens, always on access, camera, music players, faster processors, more memory—these all will consume more power. Vendors will need to develop power minimization strategies in addition to new features. The current line of products include:

- *Basic wireless phone.* The primary functions of basic wireless phones include voice calls, Caller ID, voice mail, short messaging, basic address lists, and Web browsing. The units have small displays.

FIGURE 4-2   Nokia Communicator. Courtesy of Nokia.

- *Smartphones.*   These devices are basically wireless phones with PDA-like features embedded in them. They have a separate alpha keyboard, calendar, address book, personal information management (PIM), and color screen. An example is the Nokia 9210 Communicator (Figure 4-2).
- *Two-way pagers.*   The primary benefits of two-way paging are the reliable national coverage, strong in-building reception, and long battery life compared to cellular phones. The major drawback of the messaging-only pagers is that their Web browsing capabilities are much less advanced than those of smartphones, pocket PCs, and PDAs. Examples include the RIM Blackberry and Motorola.
- *PDAs/pocket PCs (with embedded or external modems).* Personal organizers, many of which now come with a color screen and have the ability to use Windows applications such as Excel and Word. These include Palm, Compaq iPAQ, Casio, Handspring, and Hewlett Packard.

(See Chapter 2 for more detailed information on operating systems like Palm, Symbian/Epoc, WinCE, and J2ME—a version of Java.)

MOBILE PORTALS

Now that devices are moving towards an operating system that makes it easy to download and run applications designed for small mobile devices, users need a point of entry to the Internet.

A *mobile portal* is such a gateway or entry point, adapted to the particular circumstances of wireless mobile access to the Internet. In addition to optimizing access to the Wireless Internet, mobile portals aggregate and structure content and links and provide navigation tools. Mobile portals provide links to applications that can be purchased and downloaded onto the device and used on- or offline. Although Internet portals have existed for some time, these existing fixed-line portals are designed for fixed wireless access; they have content and revenue strategies that do not easily fit the needs of the mobile user.

Fixed line portals have brand recognition and access to content deals but no real experience with wireless. Wireless operators have experience in dealing with wireless voice users but know little about the portal business. The mobile portal start-ups often have experience in both areas, but lack brand recognition and financial resources. Despite these limitations the start-ups do have an advantage because they are focused on the market and not distracted by nonmobile operations.

The revenue model for mobile portals is a combination of subscription, advertising, and transaction revenue. Portal operators that have an existing billing relationship with end users currently have an advantage in terms of billing but all portal operators will eventually offer "mobile wallet" services that enable users to pay for items (including services and subscriptions) over the same wireless connections. The mobile wallet is a password-protected area in your phone which contains your credit card or debit card information. When you want to purchase something, all you need to do is retrieve the virtual credit card to complete the purchase.

Users will not be easily convinced to pay for access to content that is freely available on a fixed Internet connection. Content value must come from filtered and analyzed information as well as from secure and personalized services that maximize end user communications while minimizing time spent navigating and searching.

The mobile portals of today allow network operators to fine-tune services and applications to meet the requirements of their customers, allow operators to gain control over content, and position the customer base to be migrated to Wireless Internet applications, content, and portals, while creating entry barriers for other competitors.

MESSAGING—THE FIRST WIRELESS INTERNET APPLICATION

If you are reading this book you probably already know that short messaging service (SMS) or text messaging is getting to be a pretty hot topic these days. Everyone has heard about the SMS explosion in Europe and Asia—5 billion in March 2000, 9 billion in September 2000, and over 53 billion in the first 3 months of 2001. Amazing isn't it? Then again—short messaging started a long time ago, before digital phones or cellular even! Back when all they had was an alphabet and two towers—or two hilltops to be exact.

The Greeks invented the first wireless text messaging before the birth of electricity. Back around 400 BC the Greeks developed a way to represent each letter in the Greek alphabet using a combination of five torches, lit and unlit, to spell out messages from one hilltop to another.

It wasn't perfect—bad spelling, slow throughput, "wind static," and size were issues. And in the end it was much like today: They still didn't have great coverage in the valley and wearing the equipment on your belt made you look like a geek.

Wireless technology has allowed people all over the world to maintain business and social connections regardless of location. This emphasis on personal communications will carry over into Wireless Internet applications through various forms of messaging. Messaging applications allow users to send short pieces of information to others in near real time but allow the recipient to reply when convenient. Messaging applications will also follow the evolution from simple text to full multimedia, and in the process teach consumers a new way of communication. Messaging applications will be used in ways that simple voice communications have not been used and such applications are critical to the success of the Wireless Internet.

The initial demand for Wireless Internet usage comes from young users. Simple text messaging has been the first real wireless data success story for European and Asian carriers. Those of you who think that this was a well-planned and orchestrated strategy might be interested to know that it didn't quite roll out that way. The first SMS was sent from a PC to a GSM phone back in December 1992. Even though it was available in the majority of GSM networks and devices soon after, SMS did not actually take off until almost 1999. The challenges with simple text messaging were very similar to the challenges Wireless Internet applications face today.

GSM carriers and device manufacturers have evolved SMS over the years to overcome challenges in:

· Screen size
· Navigation
· Text input
· Billing
· Interconnection
· Roaming

Although SMS is now a primary source of revenue growth for many GSM carriers worldwide it is an application still very much in its infancy. Carriers in the Americas, especially TDMA and CDMA carriers, have been slow to roll out simple text messaging services within their own networks. Closed networks, devices not capable of originating messages, and billing issues have hindered the critical mass needed for mass adoption, but progress is being made. Most U.S. carriers have launched simple text messaging and are working towards promotions to encourage trial and usage. Interconnection among competing carrier networks is in discussion, and companies offering data clearinghouses for SMS have opened for business.

SMS, as a first Wireless Internet application, offers consumers more than voice can. Concerned carriers are discovering that SMS doesn't cannibalize voice usage, despite being an alternative, but actually supplements and may increase voice traffic.

FIGURE 4-3   RIM Blackberry. Courtesy of Research In Motion, LTD.

When compared to voice, text messaging is described as being less intrusive, more accurate, and more private as well as being fun, addictive, and lower cost than making a wireless phone call. Users often send text messages to communicate things that they would not or could not in a voice call.

The initial demand for Wireless Internet in North America came from the mobile professional and business user. The lack of simple text messaging on cellular phones created a market for simple text messaging devices such as the RIM BlackBerry device (Figure 4-3) that allows users to send and receive short email messages. The initial high cost of these devices and services has resulted in quicker adoption by business enterprise rather than the consumer youth adoption that is occurring outside of North America.

Whether we start with young consumers or business professionals, growth will depend on efforts to further educate the market about the possibilities of Wireless Internet services, applications, and content.

PERSONALIZATION

Many of these applications and services are increasingly being tailored to individual tastes. As people customize services to suit their use of the Internet, its usefulness will increase considerably. The wireless device tends to be a personal device with a single primary user, unlike personal computers that are commonly shared among users. Because mobile devices are frequently used and almost always with the user, wireless operators can exploit the wireless device's access to time- and location-critical information. As these services, content, and applications become more personalized and location based, users have richer, more rewarding, and more relevant experiences. This information will be based on the user's own information—"my news," "my banking," "my investment portfolio"—to make it absolutely specific and relevant to the user.

Additionally, every user will be able to create his own content including video, animation, still images, and text and all the information will be in digital, transferable form. User-generated content will be a very important portion of the multimedia Wireless Internet business model. The importance of the Wireless Internet device as an instrument for information, entertainment, and transactions will increase as physical boundaries dissolve.

CONTENT DELIVERY

Content delivery involves the transport of information from a source (content provider) to its destination (customer). The customer usually selects to receive content (such as travel directions or flight status information). The service provider may charge a fee for access or may receive a percentage of the fee paid by the recipient to the content provider (royalty fee). Some of the more popular content providers include mapping companies (for directions), music, flight status, weather information, and other real-time or near real–time information sources. The actual information content is often provided through an application service provider (ASP) and transferred through an Internet portal (gateway). The ASP usually manages and updates the content, and the wireless provider provides the transport to the end customer.

PERSONALIZED COMMUNICATIONS

Personalized communications consist of applications and services that are based on access to and manipulation of the user's personal data. This includes services such as personal information management, calendar and scheduler management, email messaging, unified messaging, chat, and community participation.

Wireless Internet applications will add value to personalized communications by increasing a user's ability to access personal data while mobile. We've all experienced situations where some small piece of data isn't there when we need it. But no matter how hard we wish we had not forgotten that contact name, phone number, date, or account number, it still refuses to magically appear. Wireless applications will enable users to wirelessly retrieve data that may be typically stored in various other digital systems. These applications will often be a substitute for another method of access but will add value by retrieving just enough data to get the task done.

Key drivers for personalized communication applications are:

- *Time sensitive data.* Data that has a very short useful life—wait too long and you'll miss out on some opportunity.
- *Security and privacy.* Data that you might not want to carry a hardcopy of for fear of losing it—account numbers, passwords, and personal information that you'd rather not carry with you.
- *Access to others.* Applications that allow users to contact and receive messages from others regardless of the method used by the sender—emails, faxes, voice mail, instant messages, electronic reminders, and other personal communication.

Other application areas of personal communications where mobility will add value include the following:

MOBILE ELECTRONIC MAIL

Electronic mail (email) is the transferring of information messages via an electronic communications system. Initial versions of email could send short text messages of 1 to 3 pages. Email technology has evolved (standardized) to allow file attachments, and new versions of email (such as those using Flash technology) send animation or video clips as email messages.

Email messaging is probably the best single reason for users to get connected to the Internet. There were over 400 million email account users in 1998, and the number of email accounts is expected to top 1 billion by the end of 2000.

Email messaging has been the leading application ("killer application") among online users age 18 and younger. Email is used by greater than 40 percent of online children under age 13, and almost 60 percent of online children between ages 13 and 18. A large proportion of older children spend their time online communicating with others via instant messaging.

Wireless email will grow quickly as society adapts to email as a more vital lifeline of communications, and especially as people begin to appreciate the convenience and freedom of being able to connect from anywhere. Additionally, as the wireless email landscape matures, advanced capabilities such as voice-enabled text-to-speech, real-time synchronization with desktop and calendar, intelligent filtering, and security will make wireless email services a "need to have" rather than "want to have."

INSTANT MESSAGING

Instant messaging (IM) is a very popular fixed Internet application that allows users to identify who is available for the purpose of exchanging text messages. IM is different from email or SMS in that users are able to see "presence" information.

THREE KEY BENEFITS OF WIRELESS IM APPLICATIONS
· *Interconnection to other online devices.* American wireless users are only now starting to use phones that are capable of two-way text messaging. One problem is that there won't be many other users to exchange text messages with until more users have a newer SMS-capable phone. Connecting online PC applications to wireless systems gives the wireless text users an existing embedded base of PC-based IM users with which to trade two-way text messages.

IM allows the wireless user to send and receive text messages to any person logged into the IM system with either a wireless device or PC. The online population of global IM users is about 130 million, which increases the size and value of the network.

Wireless IM requires the user to have the ability to connect to an IM system. This connection can occur over a number of different transports such as SMS, circuit-switched data, or packet-based connections. Some of these methods require the carrier to have or connect to an IM system that routes messages between PCs and wireless devices and provides presence information to users.

· *A continuous user interface.* Text messaging with SMS can be very useful for sending short messages back and forth but the effort required to open and close the application when sending more than one or two messages to the same recipient in a short time can be cumbersome. One way to improve the SMS process is via a continuous user interface that can save keystrokes and provide a better user experience. This simply means that the screen scrolls the text as the two users send each other messages. This eliminates the need to repeat the process of opening, closing, and addressing messages to the same person for the duration of the text discussion.

· *Presence information.* The majority of the value of IM over simple SMS lies in the ability to utilize presence information. Presence information is simply the ability to know who is "present" and able to chat and who is not "present" and therefore not available. Availability is key to making instant messaging "instant." The "instant" comes largely from being able to identify who is available to chat and not waste time in sending messages to those who are not.

Have you ever sent or received an email that required action right away? Getting a time-sensitive message such as "We are going to go get lunch, do you want to join us?" offers both sender and receiver little value if the message is not received, and replied to or acted on, in short order.

Although, IM is currently only used for text messaging it will evolve beyond text very quickly. Text is a form of data.

Future networks will be able to handle higher amounts of data, enough to allow voice calls over data channels as well as other media types—pictures, video, audio files, and more.

End users will rely on presence data to control and filter whom they attempt to contact and who is able to contact them based on stated availability and user-controlled profiles. Users will be able to tell the IM system they are available but also further define availability in any number of ways—available for work-related contacts only or only available after 5 p.m. for chats about happy-hour plans. This ability to control and alter a user's profile limits incoming messages to those that the user deems currently relevant and useful.

Imagine using presence info the next time you want to call someone. Often you wouldn't bother calling if you knew they weren't available; you would simply call later or perhaps choose to call someone else.

You know that those who call you or send you messages aren't all your friends or family—businesses such as retailers are eager to reach out and contact you. Technologies such as Bluetooth will allow businesses to send information to your phone as you pass near their location. This initially sounds good *if* you want this info but a real inconvenience if you don't. What if you could control what you receive? Imagine a profile that you set up to allow info from restaurants—menus, specials, seating availability—on a Friday night as you are walking around looking for a place to eat.

This could be useful. The profile would block messages from stores and other businesses that you aren't currently interested in. This could be reversed the next morning when you are in shopping mode and could care less about finding a place to eat. The ability to receive information and begin a text or voice discussion with businesses *you* choose is valuable to both parties.

You benefit by getting access to useful data when and how you choose with complete control over the filtering process. Companies benefit by knowing who is truly interested and not offending uninterested consumers with unsolicited offers. Presence information is an important tool for location-based services and other messaging services that value the ability to target who is available and potentially interested in the message.

ELECTRONIC CALENDAR

Electronic calendars are slowly gaining in popularity as connectivity improves and devices are able to synchronize data between PCs and mobile devices. Paper calendars have given way to Palm Pilots that can share and synchronize data between the device and the PC. This works great when both devices have the most up-to-date data but that's not always the case. Calendars are often workgroup productivity tools that allow groups to schedule meetings across multiple calendars based on the current data. This becomes an issue when you are away from your desk and enter a new appointment into your Palm Pilot. Before you are able to synchronize this data with the group calendar application it's possible that someone will schedule a conflicting meeting for the same day and time. Wireless calendar access will allow you to see an up-to-date schedule as well as record new entries on the network for all to see.

FIGURE 4-4   Electronic calendar. Courtesy of Compaq.

MOBILE ELECTRONIC WALLET

We all have credit and debit cards that we use for various purchases; some of us even make online purchases at home in front of the PC. When it comes to inputting these numbers into a wireless device every time we wish to make an electronic transaction most of use would rather choose an easier method of payment. Wireless mobile wallets will allow users to access account data that is stored in the device or network to simplify and ease the check-out pain by remembering credit or debit card information. The mobile wallet automatically reads the merchants check-out form and inserts user information into the appropriate fields. In the future more banks and merchants will experiment with mobile electronic wallets.

IDENTIFICATION OR SECURITY ACCESS

Whether you work in a large or small company or have a membership to a gym or country club, security access badges have become part of our personal wardrobe. These access cards typically identify who we are and give us access to areas that are limited to authorized persons only. Wireless applications will soon be able to store these same permissions and offer electronic identification with an extra touch of security. Badges are typically passive devices that only need to be swiped or passed over a reader. Security applications in wireless devices will have the ability to require users to enter passwords to activate the access ID. Wireless also enables lost access devices to be recovered easily. Lost wireless devices that are used for security or identification purposes may also have GPS location technology that pinpoints the exact location of the lost or stolen device or simply confirms that the user has not entered an area that is off-limits.

ELECTRONIC PHOTO ALBUM

A picture may be worth a thousand words, especially if you're on the phone trying to describe something. Wireless applications will allow users to carry or access images that are stored locally on the device or on a remote server. The number of photos most consumers are able to carry with them is limited by the size and fragile nature of printed photos, but digital photos take up little space and eliminate fear of loss because copies are easily made.

Since the mid-1990s, low-cost digital cameras are available that allow customers to capture and transmit digital photographs. Because digital cameras allow the customer to manipulate the digitized photos, they can be enhanced to remove red-eye, aligned, and unwanted areas can be cut out. These images can be used to create electronic postcards or greeting cards. Telecommunication network high-speed data transfer, combined with store-and-forward service capabilities, will allow customers to transmit and receive high-quality photographs. Furthermore, many wireless devices already have the ability to attach a camera to their data port.

Digital camera revenue is expected to surpass that of film cameras in 2000 for the first time ever, with $1.9 billion worth of digital cameras sold in the United States. Digital camera unit sales are expected to grow from 6.7 million in 2002 to over 42 million in 2005.[*] Figure 4-5 shows a digital camera attachment offered by Ericsson that can be connected to a phone to transmit pictures by email.

FIGURE 4-5   A combination of phone and camera attachment. Courtesy of Ericsson.

[*]Info Trends report.

NEWS AND INFORMATION

Information-based horizontal applications include directories and guides, which users can access and customize according to their interests, such as headline news, business news, specific company news tracking, weather, sports results, or stock information. Wireless Internet applications will add value to news and information by increasing users' ability to access up-to-the-minute time-sensitive news while mobile. The one thing we can all agree about is that if you wait too long, it's no longer news.

Wireless applications will enable users to wirelessly access condensed versions of news and information that is relevant for that particular moment in time. This access will often supplement other methods for accessing this data but will add value by providing timely access and notification of important events.

Key drivers for news and information applications are:

- *Time sensitive data.* Unpredictable news or events that the user has an interest in knowing about as they occur. Weather, sports scores, and news headlines are all examples of time-sensitive news.
- *Access to up to date directories and guides to ensure efficient actions.* That paper phone book won't always tell you that your client recently moved his office to a new location across town!
- *Filtered access.* News and information can be condensed and filtered based on pre-existing rules and profiles according to the user's interests.

Some of the application areas for news and information where wireless mobility will add value include:

- *Virtual newspapers and magazines.* Virtual newspapers and magazines use communication technology to deliver periodical and advertising information. In 1998, over 80 percent of consumers surveyed said they believe that the Internet is as reliable as offline (e.g., printed and television) media sources.

Because of the proliferation of 24-hour cable news channels and the increase in online news services, average daily newspaper readership fell to only 58 percent of the United States population in 1997. This is compared to over 80 percent in 1964. With only 31 percent of the 21- to 35-year-old age group reading the newspaper, traditional newspapers and broadcasters are using virtual newspapers and magazines via Internet to reach a more affluent, younger, demographic online audience.*

Newspapers rarely duplicate themselves word for word online, but they often provide more than enough for the reader without the paper edition. When viewing online newspapers, readers are not limited to selections of local newspapers. They have access to newspapers around the globe. Almost all newspapers have an online version. Additionally, the online versions are generally free (advertiser supported) and are available before the paper ones hit the stands. Online newspapers and magazines tend to offer expanded coverage into areas such a travel, entertainment, and culture. They provide exclusive content such as breaking news, live sports coverage, online shopping, opinion polls, and discussion groups. However, probably the best advantage of online newspapers is that they provide advanced search and retrieve archives to the customer. With increased available bandwidth, virtual newspapers can take advantage of video and audio media to add value to their news services.

- *Virtual or E-Books.* Virtual books or electronic books (e-books) are books in digital form that can be displayed and navigated through by a user. Many virtual books are available through personal computers or personal digital assistants (PDAs) via CD ROM or a connection to the Internet. Portable e-book readers come with leather covers, a built-in modem, and color screen.

Since 1998, online publishing offered electronic books in PostScript Descriptor File (PDF) format. E-books offer book publishers a way to control distribution if they're able to tie

---

* Newspaper Association of America; Pew Research Center.

content to a specific device. In 1999, the total U.S. book market was approximately $21 billion and the e-books market share was less than 1 percent,* due in part to poor display devices, lack of compelling content, and the limitations of the user's experience with the display device (which is far outstripped by those of digital music users).

With the introduction of better display devices and more content available via the Internet, the marketplace for virtual books should dramatically increase. It is likely that e-book vendors will focus initially on vertical opportunity segments with time-sensitive content, such as mobile maintenance (service instructions), education (distance learning), healthcare (telemedicine), and law (case histories) topics.

ENTERTAINMENT AND LIFESTYLE

This category covers applications that are primarily designed for leisure and entertainment, such as music and video-movies, horoscopes, jokes, and soap opera updates. Games, sporting events, icons, ring-tones, postcards, and video clips are included in this category. Most of these will be multifunctional, provide information and advertising, and may change according to season or nature of the event.

Entertainment and lifestyle will be a popular category of applications as users look for ways to personalize devices. These applications will offer truly personal services that follow the user throughout the day and offer bursts of entertainment when convenient.

Key drivers for entertainment and lifestyle applications are:

· *Cool, current, and compelling data.* Entertainment is a very personal and fickle subject because user preferences and content that is "in" changes often.

· *Notifications of events with limited timelines.* Concerts often sell out within minutes of being announced, leisure events are often subject to last minute promotions, and special offers make timeliness and mobility a benefit.

* Interview, L. Harte, President APDG Publishing, 21 April 2000.

• *Changes and delay*. Weather can cancel or delay travel or other personal activities with little notice. Consumers with the right wireless applications can regroup and plan an alternate course of action to better utilize personal time and efforts.

During 1999, over 19 million people worldwide used mobile phones to download or access online games, audio, or video services. Entertainment will be one of the leading forms of content carried over wireless networks. Surveys of industry confidence indicate that entertainment is perceived to be the second most popular mobile application after email and SMS.*

In 2000, simple embedded games and ring-tone downloads are popular. As new low-cost broadband wireless services become available, we will see more new applications. These include playing interactive mobile games, listening to music downloads (in MP3 format) via the wireless phone or attached accessory, and watching video clips (e.g., football highlights) on your wireless video-phone.

TRAVEL.   In 1999, US consumers booked $6.5 billion of leisure and unmanaged business travel online, almost triple the $2.2 billion booked in 1998, representing 5 percent of total US bookings in 1999. Online bookings are expected to increase significantly to 14 percent of total bookings by 2005 ($28 billion), with key segments including lodging, cruise, tour, and rental car products.†

MUSIC AND MUSIC CONTENT.   Music content delivery involves the transporting of music content (usually in digital form) from a manager of the content (a music producer or their agent) to the end customer. In the 1990s, much of the content was sold via the Internet rather than delivered through it due to the limited amount of bandwidth and devices to store and play downloaded music content. Downloading a full-length CD, even in compressed form, is a formidable challenge for the

---

* Jupiter Communications.

† Ovum, 2000.

average user with a dial-up modem. The market for digital distribution of music in 2002–2003 is estimated at approximately $150 million. With 3G broadband wireless data, it will be possible to download entire music CDs in less than 2 minutes.*

The sale of compact discs (CDs) and tapes via online services is expected to grow to $2.6 billion (14 percent of total U.S. music sales of $18.4 billion by 2003). Online shopping allows customers to easily preview content or details of a product such as tracks on music albums. In 1998, music industry revenue topped $13.5 billion in the United States, with online sales totaling $157 million, up 315 percent from 1997's figure of $37 million. As 3G wireless networks and other broadband systems are deployed, consumers will shift their acquisition from purchasing CDs or tapes to downloading their favorite music content to their media player.

By 2000, more than half of the users on the Internet had listened to music audio on a personal computer (PC). Of these, 36 percent have downloadable music and 5 percent have transferred unauthorized (pirated) music files to their hard disk drive.

As an interim approach to music content delivery on the Internet, companies are offering digitally compressed music in MP3 form. In 2000, MP3.com launched subscription music channels on the Internet. For a monthly fee of less than $10, users have access to thousands of music tracks to listen to.

GAMBLING.   Online gambling is the interactive process of allowing customers to wager money or credits in return for games that have standardized odds. Online gambling has the potential to be one of the largest interactive services. In 2000, the global gambling market was valued at over $900 billion. A growing portion of the gambling industry is moving towards online gambling. Customers with a credit card and an Internet connection are able to gamble on casino games, lotteries, and sports books (horse and dog racing, boxing, team sports betting, etc.) almost anywhere in the world.

Although there are some issues about the legality of gambling online, the majority of online gamblers are located in

* Jupiter Communications; The Recording Industry Association of America (RIAA).

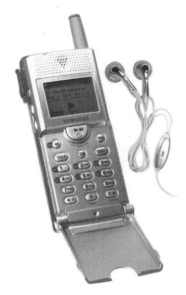

FIGURE 4-6 MP3 player/phone. Courtesy of Samsung.

countries with regulations favorable to online gambling companies. Many of these companies operate in places such as the Caribbean, Europe, Australia, and South Africa. It is projected that over $10 billion will be gambled online by 2002 as operators take advantage of the huge audience reach and cost savings of the Internet.*

NETWORKED GAMES. Since 1997, networked games have become a big opportunity on the Internet. Networked games allow users to play games against friends who are connected to the Internet. Almost any computer game that can be played by two or more people can be played online. It is estimated that by 2002, 60 percent of children online (over 16 million) will be playing games, and they will spend over $70 per year for game services. Adult will spend $140 per year for an estimated total of $622 million for online game services. Wireless high-bandwidth services allow for substantially improved game services through streaming video and audio, and permit its players to engage in games virtually from anywhere.

* Data monitor.

And as low-cost home broadband services and equipment become more available, companies will create richer gaming applications. Major game vendors such as Nintendo, Sony, and Sega are entering the broadband market by selling game CDs and allowing potential customers to participate in online games for free. Alternatively, there are some games that can only be played online including Ultima, Starseige, Quake Arena, and Unreal Tournament.

To use online games, customers pay a monthly access fee or pay-per-play. Networked games make it much easier for customers to find new opponents, or to find a partner to play at any time. High-speed data wireless access will provide for much better three-dimensional (3D) graphics viewing.

VIDEO ACCESS AND MOVIE RENTAL.   Video content delivery will be one of the leading drivers of the 3G broadband marketplace. Consumers have a voracious appetite for all types of media, particularly video (movie) content. In 1999, over 70 percent of households in the United States rented an average of 1.3 videos per week.* The statistics for movie rentals confirm the preference of movie viewers to stay at home to view movie content. Since 1980, when VCRs first emerged as a means of watching full-length motion pictures, the sales of pre-recorded rental and sell-through video cassettes has grown by more than 66,000 percent as compared to box office theater growth of 22 percent over the same period.† The video rental business is projected to top $7 billion in 2000 and grow to $19 billion by 2004, with video sales reaching $20 billion.‡

Adult entertainment content ordering and delivery has been one of the leading categories of early Internet usage. As such, adult entertainment was an early adopter of user interface augmentation through streaming video, private access to sensitive material, and one-click ordering. In 1998, pay-per-view and subscription adult entertainment accounted for about 40 percent of the U.S. consumer paid online content market.§ Adult

* Video Store Magazine, January 1999.
† Motion Picture Association.
‡Paul Kagan Associates.
§ Jupiter Communications.

entertainment, a multibillion dollar industry, will benefit from broadband access. Consumers will be able to download private content to their wireless devices or wireless-enabled viewers.

VIRTUAL RADIO STATIONS.   Virtual radio stations are digital audio sources connected to a network (typically the Internet). In 1999, there were over 2,000 radio stations operating on the Internet. Virtual radio stations have a strong competitive advantage compared to standard radio broadcasts. Radio stations Web sites can do more than simply rebroadcast their on-air signals. They can provide photos of disc jockeys, contest prizes, and winners, and act as current news centers for entertainment events and weather services.

Broadcast radio stations have been offering content delivery by both radio and Web access in anticipation of a significant shift to Internet (virtual) radio. Internet radio offers the ability to customize (personalize) a broadcast to groups or individual receivers. By 2005, 41 percent of the population will listen to personalized, on-demand audio content at least once a week. Content providers will adopt genre-specific business models.

Radio stations are taking aggressive steps in developing a new breed of Web sites designed to offer fresh content and help the media outlets connect better with their target audience. This includes offering chat rooms, news updates and music reviews, and other social-based services that make their Web sites more appealing. Additionally, virtual radio stations can use their Web sites as research tools to determine listener preferences. The system serves up real-time information, providing details on the music being played. Listeners then are asked to use the Web site to vote on the song being played, thus giving station programmers instant feedback on listener tastes. The radio stations then talk up their Web sites during radio broadcasting, driving more usage to the Internet service. The Internet is having both a positive and negative impact on radio station ratings and revenues.*

* www.electricvillage.com.

FIGURE 4-7   FM radio module attachment for a phone. Courtesy of Ericsson.

VIRTUAL TELEVISION STATIONS.   Virtual television stations distribute digital video and audio through the Internet to groups of viewers. With broadband digital video access, the Internet will become a new avenue of distribution for broadcasters that hope to target previously unreachable mobile audiences.

Since 1999, there has been growing public interest in interactive TV (iTV). This has been led by satellite and cable systems deploying subscriber equipment and infrastructure capable of delivering a variety of interactive services. Some of these early interactive functions include an electronic program guide (EPG) and parental control through channel-locking features. A type of one-way datacasting on virtual television stations allows viewers to choose from limited, primarily text-based, supplementary content.

Other virtual television features and functions may include game or quiz show audience participation. These features and functions all present new opportunities as well as challenges to programmers, advertisers, and providers of interactive services as they navigate through a maze of complex platform landscapes defined by a complicated mix of networks, set-top boxes, and software. It's projected that 35 percent of U.S. households (over 25 million homes) will use some form of interactive TV services by the end of 2005.*

IMAGE AND VIDEO PRODUCTION.   Images and video can be captured in electronic form and transferred to other locations.

* Jupiter Communications.

FIGURE 4-8   TV phone. Courtesy of Samsung.

Because of the large file size of high-resolution images, media transfer has primarily been in the form of high-density disks or video tape.

Although initial production of images or video is in a studio, the production of edited images, video segments, or computer animation may be performed at many different locations. Broadband connections allow for editors and producers to interconnect without the delay of shipping storage media.

INTERACTIVE TOYS.   Interactive toys will utilize wireless communication technology to better interact with other toys and to update software programs. Wireless updates will allow toys to keep current with a player's abilities or interests while increasing the useful life of the device. Interactive toys have motors, sensors, and infrared messaging and speech recognition technologies that respond to communication signals and originate messages. The responses may be in the form of mechanical action or an audio message.

Interactive toys have been available for many years. Some of the first interactive toys responded to signals that were sent via a television channel. These toys responded to colors or patterns within the television signal.

Interactive technologies, specifically the user interface, are constantly improving. With wireless connectivity for example, interactive toys such as virtual pets with artificial lives, will become more "alive" thus offering simultaneous physical, verbal, and PC-like interaction virtually anywhere.

LOCATION-BASED SERVICES

Navigation and tracking are practical services that will be highly valued by users, in that all these same services can also be linked to advertising and commerce to provide, for example, directions to the nearest Italian restaurant, or to a store with current sales promotions.

Location-based services will offer value to applications that benefit from geographically filtered information. These applications can eliminate the effort required to locate or request data by first determining what data might be relevant to the user's location. Knowing that I am currently in a mall would be useful if I am searching for listings of shoe stores—the application can first serve up those shoe stores that are located in the mall before offering those that are across town.

Key drivers for location-based service applications are:

· *Ability to filter content based on geographic relevance.* Information, places, or listings of events that are close to a user's current location. Special information can be in the form of alerts to hazards or situations that can impact the end user.

· *Ability to identify and transmit location.* When users are lost or otherwise unable to communicate, such as after an accident or injury.

· *Community-building possibilities.* Applications that utilize the end user's location to match them with others interested in the same activity or event including dating, finding sports activity partners, and the like.

FIGURE 4-9   Nav Talk, integrated cellular phone with GPS. Courtesy of Garmin

TARGETED ADVERTISING.   Targeted advertising is the customizing and individual tracking of advertising to the specific recipient of the advertisement. Wireless Internet systems can customize, deliver, and track multimedia advertising to specific groups of individuals.

Advertising is traditionally associated with the promotion of branded goods and services. Because of the intolerance issues of users in the wireless environment such as paying for incoming calls and airtime privacy invasion, advertising should be positioned differently with different associations. Advertising over wireless should be linked to content, location, and e-commerce, which will enable advertising to be positioned as a useful service.

ENTERPRISE LOGISTICS.   Wireless networks offer the ability to track the position of mobile phones (within a limited distance) and provide information services based on the determination of the location. Navigation and tracking service are highly valued by users in key market sectors such as truck dispatch management and public vehicle management (e.g., buses).

Access and Connectivity-Enabled Solutions

This category includes both horizontal and vertical applications and services that are enhanced by wireless. The applications and services described in this section will often make use of portable computers such as laptops, which are able to provide value primarily due to connections in locations where connections are not easily obtained. A wide range of industries will benefit from wireless connectivity and services; these include intranet and company database access, plus mobile office applications such as file transfer and workgroup applications. The key vertical applications in this category include operational and automation applications, such as sales order entry and dispatch. Other integrated vertical applications may serve specific sectors in a growing number of consumer, education, and healthcare markets.

Key drivers for business-oriented applications and services are:

· Basic connectivity to the Internet or corporate network—access to more than specific applications.
· Ability to set up and install communications access quickly and in areas where fixed line access is not readily available.
· Enabling of remote access while mobile.

Some of the application areas where wireless mobility will add value include the following:

Business Applications. Mobile office and intranet (internal company Internet) applications are essential for business and corporate users. Business users are usually the highest spending and highest usage market segments.

The initial demand for wireless services will likely be generated by the business and vertical (specific industry applications that solve a business problem) sectors, because business customers have the greatest need for the high-speed services. These services need to be time-critical (existing business users are frustrated by the slow speed of cellular data). It is also easier to justify a financial benefit for business users compared to residential (entertainment) users.

Business users have the greatest need for applications such as file transfer or email with attachments using mobile (location independent) delivery. These applications could be significantly improved by the increase in data speed as offered by 3G wireless systems. Mobile data applications are already being used in specific types of companies, such as utilities, to operate and maintain critical facilities. Companies that use mobile data for these applications are committed heavy users.

VIDEO CONFERENCING. Video conferencing combines dedicated audio, video, and communication networking technology for real-time interaction. Companies use video conferencing to reduce or eliminate travel while allowing employees to interact.

New applications such as Microsoft's NetMeeting, offer conference attendees at two (or more) locations real-time voice and video conferencing. In addition, many video conferencing applications include collaborative application sharing (for shared presentations), multiperson document editing, background file transfer, and a whiteboard (real-time shared interactive displays) to draw and paste on. The projected lower cost and high-bandwidth capability of next generation wireless systems will allow more cost effective and portable video conferencing services.

Figure 4-10 shows a wireless video phone product concept by Nokia. This video phone can both send and receive images via high-speed wireless systems.

FIGURE 4-10   Wireless video phone. Courtesy of Nokia.

REMOTE CORPORATE NETWORK CONNECTIONS. Remote corporate network connections allow company employees to access company networks and receive services (e.g., rapid file transfer) as they would experience if they were located (working) at the corporation. The rise in "virtual corporations" has resulted from increased worker productivity, reduced facilities costs, and satisfaction of environmental and regulatory requirements for reduced number of commuters. It is estimated that over 7 percent of all workers in the United States spend at least some or all of their time as teleworkers. This growth in the home-based work environment has been a major driver for home and business network interconnection speed.

BUSINESS KIOSKS. Business kiosks are remote locations for business retail centers. Business kiosks may be unmanned or satellite offices that require connection to a head office or stand-alone information centers that require periodic information updating.

The use of business kiosks allows companies to expand their market territories without significant risk or capital investment. By utilizing wireless data connectivity, kiosks can be installed quickly and at low cost.

Public Internet Kiosks are a type of pay phone booth that contains a computer terminal that can access the Internet. For a nominal price, a customer can check email or browse the Internet. Most public Internet kiosks are scattered throughout public places such as airports, train stations, convention centers, hotels, office building lobbies, and shopping malls. These public Internet kiosks can be used as a media center for information services.

Internet kiosks can be multipurpose or adapted to satisfy specific needs. They can be used as automated teller machines, travel service providers, ticket centers, and to provide other business services.

In 1998, there were approximately 10,000 kiosks in the United States, and the number is expected to rise to more than 100,000 by 2002. The typical cost of a kiosk is $35,000 to $55,000, in addition to monthly space rental fees.*

* Summit Report.

CUSTOMER CARE.   Customer care is the process of answering customer questions about a company's products or services. It is estimated that over 65 percent of the cost of providing customer support service originates from simple product and billing questions.*

The cost of customer service is greatly reduced and customer satisfaction is dramatically improved as customers and suppliers are able to satisfy their information need via the Internet. Furthermore, the capabilities offered by 3G wireless Internet provide for even greater flexibility and convenience from the field. The information gathered from the areas regularly visited by a browsing customer allows companies to promote similar products and services to them.

DOCUMENTATION MANAGEMENT.   Documentation management includes the capture, storage, organization, and coordination of access to large amounts of text and image information. This information may be stored at one or more locations and the information may be accessed or transferred to display devices (terminals), printers, or other repositories (for long-term storage).

Documentation management allows manuals, procedures, specifications, and other vital information to be instantly accessible by authorized employees. Documentation management can save a company a considerable amount in printing reproduction costs, because all documentation is digital rather than paper.

FIELD SERVICE.   Field service personnel interact with clients or equipment in the field. This personnel has traditionally had limited access to company materials. Using 3G broadband communications systems, field service personnel can access documents (e.g., company catalogs and service manuals) and example procedures (e.g., video clips), capture information (e.g., using a digital camera to record an insurance claim), and obtain assistance in the repair of equipment (e.g., connect systems for remote diagnostics). Figure 4-11 shows a personal

* Interview, Steve Kellogg, 6 May 2000.

FIGURE 4-11   Personal digital assistant, the Nokia 9210 Communicator.
Courtesy of Nokia.

digital assistant (PDA) that allows a field service representative to access various forms of media.

MANUFACTURING

Telecommunication systems have long been used in manufacturing processes to monitor and control production to ensure quality. Manufacturing systems can benefit from wireless production monitoring and low-cost data communication systems.

Production monitoring is the process of using data devices or sensors (e.g., video cameras and keypads) that transfer information via communications lines to keep records of physical production. The Internet and other communication networks are moving onto the factory floor to provide companies with an inexpensive means to link workers and the machines they operate to remote repositories of information. Distant managers can watch what's going on, literally, from wherever they are, through sensors, tiny Web cameras, and Web displays built directly into equipment deployed on assembly lines. Previously, these monitoring devices required physical, wired connections that limited their routing to production managers. By using the Internet or other wireless technologies, managers located in distant facilities can monitor production and ensure problems can be resolved long before the problem causes lost production or injury to personnel. Software that integrates Internet technologies into factory operations was a small percentage of a $4.8 billion market in 1999. Prepackaged manufacturing monitoring software is growing by 14.2 percent a year.*

TELEMEDICINE

Telemedicine provides medical services through the assistance of telecommunications. Telemedicine does not completely replace medical expertise, but it is critical to providing quality and efficient health care services.

Telemedicine is a rapidly growing part of the medical information management market and is one of the largest and fastest growing segments of the healthcare device industry. The expected revenue by the end of 2000 is $21 billion.

In the United States, more than 60 percent of federal telemedicine projects were initiated since 1998. The concept of telemedicine exploits much of the state-of-the-art technology available, especially if it is combined with the growth of the Internet and World Wide Web (WWW).† 3G networks will further guarantee a wireless extension of Internet-based services and technologies. By reducing the time spent in copying, sending, and archiving medical information, the cost of administration and insurance claim processing is reduced. Mobile medicine will enable healthcare workers to receive supply-on-demand content in a mobile environment.

Some of the advanced telemedicine applications include telecardiology, teleradiology, and telepsychiatry. *Telecardiology* services incorporate transmission of ECG data, echocardio-

---

* International Data Corporation (IDC).

† Telemedicine Information Exchange (TIE).

grams, heart sounds and murmurs, and cardiology images, and can be performed in both store-and-forward and interactive media. *Teleradiology* is the most widely adopted of all telemedicine applications. Clinical radiology requires prompt, near real-time transmission of still-frame images, but may also demand live or full-motion video image communication and display. *Telepsychiatry* allows psychiatric care to be conducted at a distance to provide care more frequently to patients in outlying areas.

Telemedicine applications usually encompass computer, video, and telecommunications technologies—each with its own role to play in the acquisition, transport, and display of medical information. Some of the key areas related to telemedicine include patient record management and mobile clinics.

PATIENT RECORD MANAGEMENT. Patient record management involves the storage and retrieval of medical information related to a specific person. Patient information may be gathered manually (such as an X-ray on film) or electronically (such as a patient history data record). Patient record management via telemedicine involves converting nonelectronic forms of information (such as the X-ray) into electronic forms (data files) and managing these data files to integrate data, voice, digitized images, or video. These files are stored in a computer and can be transmitted to workstations at a medical center, physician's office, or other site equipped to manage the telemedicine information request. Rapidly transporting image data and diagnoses between clinicians and medical doctors can add substantially to improved patient care.

MOBILE CLINICS. Mobile clinics are transportable facilities where health care specialists can treat patients. Using wireless high-resolution video conferencing, mobile clinics in the form of buses or vans can travel throughout rural areas with clinical technicians bringing hospital-type facilities to remote areas. The clinical technician coordinates communications with medical experts via wireless video conferencing consultations.

These telemedicine videoconference facilities allow hospital-based physicians to view patient wounds from a live video image. The traditional method requires visiting nurses to take Polaroid photographs of wounds and forward them to physicians for review. From the snapshot, the physician assesses how the wound is progressing and determines whether changes in medication or treatment are needed. Using mobile medicine, visiting nurses dial the physician, forward the image in real-time, and facilitate interaction between patients and hospital-based providers. Images can be captured and stored in an electronic medical record. The technology can help reduce the cost of continuing inappropriate therapy and shorten the time between data collection and decision making.

DISTANCE LEARNING

Distance learning provides training to remote locations. Distance learning has been available for many years and can be categorized into public education (grades K–12), university and colleges, professional (industry), government training, and military training segments. In the early years, distance learning was provided through the use of books and other printed materials and was commonly referred to as *correspondence courses.*

Distance learning has evolved through the use of broadcast media (e.g., televisions) and moved on to individual or small group training through the availability of video-based training (VBT) or computer-based training (CBT). These systems evolved into interactive distance learning (IDL).

Distance learning relies on communication systems (e.g., phone lines or mail) to connect students and teacher as an alternative to classroom training. Electronic learning (eLearning) is a form of distance learning that is becoming a viable option to traditional teaching methods and is poised for major growth over the next several years.

Through the ability of broadband video and interactive graphic technologies, students are exposed to far greater education stimulus than in the traditional learning environment. Integrated sound, motion, images, and text all serve to create a rich new learning atmosphere and substantially increase student involvement in the learning process.

The rapidly changing global economy is forcing industry professionals to continually update their skill sets. Adults may change their occupations several times in a lifetime as technologies and skill sets become outdated. This requires continual learning for adults. Adults between the age of 35 and 45 are the fastest growing group of college learners.* To advance or consolidate their careers, over 5 million adults complete some form of distance learning each year.† This is one of the primary reasons why online learning is booming, especially among working adults with children. Distance learning via broadband connectivity allows adults to attend classes in the comfort of their living room or study, at their convenience.

Many online universities, including training and professional specialty course programs, are catering to the rising demand of industry to deliver skill-development courses to the desktop at remote locations. These schools offer Web-based professional certificates as well as associate and bachelor's degrees that are built around a solid core of business and computer classes. Companies rely on these certificates to ensure employees are qualified for their new jobs.

In 1999, most online classes did not require that students have the latest high-powered computer. However, they did require Internet access (via low-speed analog modem). These distance learning courses were provided using low-resolution graphics or slow-scan Web video. As broadband services become more available and cost effective, it is predicted that distance learning courses will evolve to use high-resolution services such as high-resolution video conferencing.‡ Online distance learning courses can be accredited by regional accrediting agencies or via the Distance Education and Training Council.

PUBLIC (K-12) EDUCATION.   Elementary education involves developing fundamental skills in children and young adults. Elementary education is normally funded and managed by gov-

---

* Adults 35–45 the most rapid growth education market.

† Over 5 million adults complete distance learning courses each year.

‡ Interview, industry expert, 6 May 2000.

ernment agencies. It is the goal of many public education programs to provide the same education opportunity to all the members of a society, regardless of the economic status of its students or the demographic structure of a community.

The economics of traditional public education systems limit the offering of specific courses to regions that have a sufficient density of students. To ensure that each student can be offered the same education opportunities, distance education can offer more courses to each student. Distance education also allows students to interact with other students with similar interests and needs at remote locations. Distance learning applications delivered through the Internet can provide access to standardized courses that provide equal education opportunities to most students. Additional e-books will provide students the ability to carry a single book.

By the end of 1998, approximately 89 percent of all public secondary and 76 percent of elementary schools in the United States were connected to the Internet. Since then, public schools continue to make progress toward meeting the goal of connecting every school to the Internet by the year 2000. (In 1994 only 35 percent of public schools in the United States were connected to the Internet.) In addition to having every school connected to the Information Superhighway, a second goal is to have every classroom, library, and media-lab connected to the Internet. Schools are making great strides to achieve this; and in 1998, 51 percent of instructional rooms in public schools were connected.*

Connection speed is one of the key determinants of the extent to which schools make use of the Internet. In 1998, higher speed connections using a dedicated line were used by 65 percent of public schools. Additionally, large schools with Internet access are more likely to connect using broadband access technology.†

COLLEGE AND UNIVERSITY EDUCATION. Since the Internet was pioneered at universities to facilitate information sharing, it's not surprising that an increasing number of them are cre-

---

* U.S. Department of Education; National Center for Education Statistics.

† U.S. Department of Education.

ating Web-based universities. By 2002, 85 percent of two-year colleges (in 1999 there were 847 two-year colleges in the United States) are expected to be offering distance learning courses, up from 58 percent in 1998. It is projected that over 80 percent of the four-year colleges (in 1999 there were 1,472 four-year colleges and universities in the United States) will be offering distance learning courses in 2002, up from 62 percent in 1998. Many of these will be Web-based. To put this into perspective, there are 15 million full- and part-time college students in the United States, of which an estimated 90 percent are online, representing by far the most active single group on the Net. Moreover, in 1998, 21 percent of these students purchased $900 million in goods and services online.*

It is estimated that 93 percent of distance learning programs in American colleges and universities use email and almost 60 percent use email in conjunction with the Web.†

When distance education is offered, campus visits are not required for most programs. Learners register online each semester and may take single courses for personal enrichment or opt to enter a degree program. Textbooks and class syllabi can be mailed to learners. Online classes run typically on a 16-week semester schedule, beginning and ending at the same time as on-campus classes. Students read their textbooks and visit online message boards weekly, posting class comments or questions whenever it is convenient for them. The back-and-forth commentary on the message boards simulates a classroom discussion. Midterm and final exams are usually taken under the watchful eye of an approved proctor at a local college, library, or human resources training center.

PROFESSIONAL.   Because technology and business processes are constantly changing, professional education is developed and provided by companies to keep their employees competitive. Training budgets range from 1 to 5 percent of a company's gross sales, and a growing percentage of these funds are used for distance learning courses.

---

* e-Marketer; Student Monitor LLC.

† The Survey of Distance Learning Programs in Higher Education, 1999.

GOVERNMENT.   Providing education for government workers is necessary to ensure that information-intensive systems (such as tax collection) can operate effectively. In the United States in 1999, there were more than 3 million government workers. The average government worker receives 1 to 2 weeks of training per year to learn software and technology systems, standard processes, and to develop leadership skills. This results in a requirement of over 5 million weeks of training. To minimize the costs of travel and lost time, many government agencies use distance learning programs to reduce training costs.

SECURITY VIDEO MONITORING

Security video monitoring applications help to visually assure that valuable assets are not eroded or destroyed by unauthorized users. Traditionally, security video monitoring was limited to on-site video monitors that security personnel viewed as either videotapes or as real-time images. The introduction of low-cost digital video cameras and data connections allow for the remote location of video cameras. When these cameras are connected through the Internet, they are called Web cameras (WebCams.)

At the end of 2000, there were already in excess of 100,000 public WebCams in operation throughout the world* and private video monitoring systems have millions of privately installed video cameras. Although many of these video cameras are connected by wire, some are connected by wireless links.

The key applications for wireless security monitoring included traffic management (traffic cams), public access monitoring (public safety), law enforcement (cameras on police cars), and other applications that require a camera at remote locations where wired connections are not practical or where mobility (video monitoring while moving) is important.

Figure 4-12 shows the video camera that is normally mounted in police cars. Using high-speed wireless systems, images from police cars can be monitored at a central facility. This may dramatically increase the safety for police officers.

* APDG Research, Broadband Applications, 31 December 2000.

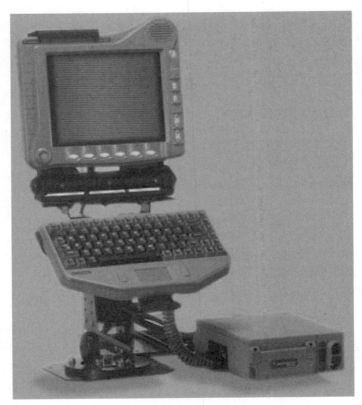

FIGURE 4-12   Mobile data terminal. Courtesy of Motorola Inc.

BILLING AND SECURITY ISSUES

Voice-based wireless devices may have started out as new and exciting technology but their success was partially due to their ability to connect to the old dull technology of landline telephones. Imagine what would have happened if early cellular phones were only able to call other cellular phones. Who would these early users have been able to call? Perhaps the only other users would have been a very small and exclusive club of executives with a habit of traveling in remote places. It's clear that cellular devices and service would not have been very useful to mainstream society. (Unless you consider the value generated by keeping the boss entertained while out in the boondocks—useful for some, I'm sure.)

The CB radio is a good example of a device that is limited to communicating only with other CB radios. CB's are relatively cheap to buy and free to talk on—no roaming or per minute usage fees. But they don't offer the best value because you can't reach everyone you wish to speak with, those you do try to reach aren't always on, and privacy doesn't exist. This points out the need for interconnections and the ability to convert from one protocol or system to another.

It's relatively easy to create concepts for cool new devices and services but regardless of how cool and unique a device or service, the value comes from the ability to connect and communicate to others.

Metcalfe's law is the theory that the value of the network increases exponentially. If you double the size of the network (let's say from a network that serves 1,000 users to one that reaches 2,000 users) you more than double the value—you actually quadruple the value of the network.

Imagine if someone were to approach you and offer to switch your Internet service to one that was half the cost. The only difference would be that this cheaper service could only reach half of your current network. Would you be willing to switch? Most likely you would not.

This is why both the telephony system and the Internet continue to grow in both users and value—the greater the size of the network the greater the value.

Dr. Robert M. Metcalfe, inventor of Ethernet, founder of 3Com Corp. and columnist for Infoworld. Dr. Metcalfe together with D.R.Boggs invented Ethernet back in 1973 while working for Xerox Corporation in their Palo Alto Research Center (PARC). PARC was also involved in the invention of the PC and Graphical User interface; however, it was not the place where these innovations were developed into successful commercial products. Ethernet, the name for Local Area Networking (LAN) technology turned PC's into communication tools by linking them together into a common network.

In 1979, Metcalfe left Xerox and founded 3Com Corp. (so named for three words—computer, communication and compatibility) for the purpose of promoting PC LAN's and Ethernet as the standard. Metcalfe was successful in bringing together companies such as Digital Equipment, Intel and Xerox and made Ethernet the most widely used LAN.

Metcalfe retired from 3Com in 1990 and embarked on a career in journalism that led to his writing a weekly column on networking, "From the Ether," for Infoworld. It was in this forum that he published "From the Ether: A network becomes more valuable as it reaches more users;" Infoworld Magazine, October 2, 1995. This article set forth the principles now known as Metcalfe's Law.

This law is very important to the growth and adoption of wireless Internet. The premise that the *network value grows by the square of the size of the network* does not discriminate by device or access method. The number of network users can be a combination of various access devices using various access technologies, each with their own network speeds. The sheer number of wireless Internet mobile devices forecasted to be in use in the future creates the yet another source of network growth and source of value for all users of the network.

METCALFE'S LAW IN REVERSE

There is a growing trend among wired and wireless services to split the Internet into isolated mini-networks and gain an advantage in directing and controlling users activities. This technique is called the "Walled Garden" approach in that the user is allowed access only to a limited area under the control of the service provider. While giving the appearance of advantage to the smaller closed network that can choose what content and services the users are able to interact with, the reality is that the mini-network loses value exponentially.

Examples of attempts to split the Internet into smaller isolated networks include:

- *The early days of email.* Early email providers resisted giving send/receive access to users of other email systems.
- *Instant messaging systems.* Closed IM systems such as AOL IM have repeatedly blocked efforts of smaller IM systems to interconnect users of various competing IM systems.
- *WAP portals.* Many wireless carriers have created WAP services that are only able to access sites offered by the carrier or carrier friendly providers. By locking the device settings or WAP server settings other sites are blocked from access even when users know and attempt to enter the competing WAP URLs.
- *Blocking or banning the ability of other sites to link to specific pages on another site (deep linking) requiring users to access content only via the original homepage links.* Some content owners want to block any traffic that does not originate from

their marketing efforts while others welcome official or un-official affiliates that point to their site.

The effect of Metcalfe's law in reverse reduces value for all parties involved. The networks users lose the value of a larger more complete network. The networks not only lose value individually but as a (now separate) group as well. Because of Metcalfe's law, the largest network always wins over smaller networks even when the smaller network initially offers value creating features or benefits. As the larger networks grow, the value of the sheer number of reachable users, services and content ultimately favors the larger networks. Since the Internet is the largest network of them all, it will not only eventually win over smaller proprietary networks (wired or wireless) but the wired internet and wireless internet will only benefit each other as they join together into a common network.

Attempts to create walled gardens of content and services for wireless Internet users have failed (and even been declared illegal) in many European and Asian countries. Service providers will only succeed in creating long-term value by focusing on quality of services instead of limiting access to other perhaps better services.

WHO CAN I CONNECT WITH?

To increase the number of contactable users available and reachable on a network requires the ability to be able to reach others on different systems with different devices using various protocols. The voice telephony system in the United States today uses something called the SS7 layer as a common method that all phone systems can use to transport voice calls from one system to another. The same common connection for Wireless data does not yet exist but is critical for growth of the wireless Internet. This interconnection—ubiquitous availability to communicate across networks, protocols, and media types—will create the critical mass of users necessary to add value to the network. Superior technology that does not interconnect to legacy systems will offer limited value due to the smaller network size and limited connections to other users who do not have the same device or service.

Some systems allow for data to be sent but do not yet offer a way to reply in kind. This inability to reply suffocates innovation and adoption of new features and systems. For example all digital phones are able to receive a text message but not all have the required software to reply. Cellular phones on one carriers system do not currently have the ability to send or receive text messages from phones on other systems.

Proprietary technology only works if it allows for interconnection. Many new features have these same issue and therefore must start out focused on a niche market—a community of some type, like a workgroup, service tech group, plumbers union, or some other closed group that can all agree to use the technology. Nextel is an example of a service that mixes general access with private group access. Nextel wireless services offer a unique service that allows the user to communicate with others in a group by simply pushing a button and speaking. This push-to-talk feature is great but only works if the others you wish to contact have the same device on the same network.

The Internet is all about access to anything and everything—the Wireless Internet must offer the same broad access but Wireless Internet data will need to be altered to fit users' needs in terms of technology, device, and environment.

UNIVERSAL MESSAGING— HAVE IT YOUR WAY

The history of voice telephony has had little differentiation— local versus long distance, similar input and output devices voice in, voice out. We spoke with our mouths and listened with our ears. A louder handset for the hard of hearing, a tape recorder for storing the "data," and eventually the ability to "share" voice with more than one person was as innovative as it got (and this hampered by the fact that no one seems to like being put on speakerphone).

The ability to visually present data changes everything: It increases the methods of input as well as allowing for various types of output depending on the receiver's current situation and preference.

Cellular phones and PDAs are improving in many ways to better accommodate the Wireless Internet. While some devices opt for a qwerty type keyboard, others have operating systems that allow the use of stylus or pointer devices to navigate and input data. All have increased display size and resolution, and many have added color as well.

It's easy to find examples of situations when voice is not the most convenient way to communicate detailed information quickly and discreetly. Imagine you are in an important meeting where the speaker has gone over the time limit, yet it would be seen as rude to leave the room. The problem is you are expected to meet your spouse at a restaurant and it looks like you will be delayed by about an hour. The information you need to communicate to your spouse is essentially:

> "Hi honey! Look, I'm still at work in a meeting that's running late and I'm not sure when this guy is going to shut up and let us get out of here. Can you please call the restaurant and move our reservation back an hour? If I can escape this snore fest and avoid my boss on the way out of the building I'm sure I can make it by then OK? Love you Buttercup!"

Given the choice to make a wireless phone call or send a text message to adjust your dinner plans, what would you do? If you were my co-worker, I sincerely hope you would choose the text method. I'm sure we can all agree that work environments can become uncomfortable when co-workers allow personal phone calls to be overheard and share too much personal information.

Mobility combined with nonvoice data creates a need for the user to choose the preferred input and output for each communication situation. Data can be manipulated in more ways than pure voice and still communicate most if not the entire intended message. Data such as text, graphics, and video can be filtered, shortened, condensed, and reformatted to better fit the output device.

Universal messaging has a history of being defined to include only what a given vendor's solution currently includes—a system that allows the user to combine voicemail and email might be called "universal messaging" even if it does not truly allow the sender to leave a voicemail that the receiver can read over email. Instead the system might send the voicemail as an audio file to the receivers' email inbox.

True universal messaging includes the ability to convert inputs such as voice and text into the output preferred by the receiver. Content that is visual—photos and video clips—would not be converted into text but may someday be summarized by an intelligent software program that could better describe the content. This is a 20-second color video clip that shows a group of children blowing out candles at a birthday party—so that the end user can choose whether to view it, save it for later, or delete it.

Possible conversions include text-to-voice, voice-to-text, email-to-fax, video-to-still images, and many other combinations, such as condensing text and then converting it to voice. Digital also adds the ability to sort, search, match, and store all types data.

OK, NOW WHO'S GOING TO PAY FOR ALL THIS?

Billing is one of the more important variables that will impact the success of a Wireless Internet. It will be increasingly important to have multiple billing models to suit a wide range of users. Users will vary demographically, geographically, by access type, and by media and content type as well as by the need to have billing information filtered or summarized. Billing not only provides access and content providers with a source of revenue but it will increasingly provide detail that helps to determine which type of content is offered and promoted.

The early days of Internet revealed its inherent difficulty for billing. The Internet was (and still is, to some degree) synonymous with "free." Users only pay for access; most content is free. It is difficult to assess value of content because users of content do not always agree about what is valuable and what is not. We can assume that those who visit a particular site receive value from it, but it is still hard to quantify that value in terms of dollar amounts.

The concept of value is—anything is worth what the market will bear. The early Internet was more like an open warehouse than a store—you found what you wanted and took as much of it as you liked. Because the content was in digital format, you could take as much as you wanted and still leave the same amount for another guy. The supply was virtually endless: A digital content "inventory" doesn't get depleted except in terms of availability via an access channel. If too many people try to take out—or download—content at once, you have the equivalent of waiting in a checkout lane. Bandwidth limitations place the only restriction on what can be downloaded from the free Internet.

Because there was no system in place to request money in exchange for content, it was difficult to determine if the content was worth one cent or one million dollars. After all, it's amazing what people will take when it's for free! Originally, this didn't bother the content-creation guys; they all thought that they would make money later, after everyone was hooked on the idea of getting content from their particular "shop." Branding was the byword, and the great World Wide Web was an open shopping mall of opportunities. Unfortunately, this didn't work too well. Giving content away cost an amazing amount of real money.

ENTER THE CASH REGISTER

Billing is not only a challenge in terms of method of payment but it is an important part of the adoption process. Billing and the pricing model used for selling content and services need to adapt to the evolving product lifecycle. The degree of experience and acceptance level of the target market calls for a varied approach when pricing new content and services. There are three basic billing/pricing philosophies:

- Free
- Flat rate (all you can eat)
- Usage based (ala carte)

FREE. Someone else pays for what the users get. Maybe its the content owners who forego payment in hopes of some other future value, or sponsors might help defray costs via advertising. This method is great for encouraging trials, but it is difficult to sell that which had been free yesterday. We've all encountered free stuff—remember those home cooked meals as a kid? What do you think would have happened if mom decided to start charging you after all these years? Can you imagine your dad charging companies to advertise to you during dinner in an attempt to defray household expenses?

FLAT RATE ACCESS, ALL YOU CAN EAT OR BUCKET PLANS. These plans offer either unlimited access or at least more access than the service provider thinks you are likely to consume. The advent of the "bucket" plan rocked the cellular industry by giving users a large enough "bucket" of voice minutes so that they became less sensitive to the time spent on wireless calls.

Price is not dependent on how much is consumed. Light users subsidize heavy users and everyone pays the same entry fee. These plans are great for heavy users who worry about going over budget and want predictable expenses. These pricing plans are bad for light or infrequent users who don't want a periodic fee for less frequent usage.

These plans have worked well for voice in large part because everyone understands how much they might use—not everyone is sure of how much they will use new data services.

For this reason, flat rate plans can be bad for encouraging a trial of new services when end users expect usage to be low. It's hard to assess how often you'll use new content or services until you have tried them for a while. But who wants to pay for the month upfront before you know? Difficulty in canceling subscriptions adds to the barrier for new services. Flat rate or bucket plans are great for encouraging the trial of new services when end users expect usage to be high or for frequent users after they become familiar with typical usage and want to control costs.

ALA CARTE OVERTURNS THE BUFFET. Usage based billing isn't only about volume but will increasingly be about quality, speed and privacy. Voice telephony has historically been incredibly consistent in terms of quality. We've not really had the need to think of what level of quality we desired when making a voice call—they were all essentially the same. The future will offer many new choices for voice and data quality as systems will be increasingly varied, each with it's own particular application and price points.

Users needs are very diverse and willingness to pay is widely varied. Similar versions of content are already billed differently in media other than the Internet: One might pay $75 to attend a concert in person, $40 for a pay-per-view showing on TV, $30 for a recorded videotape, $20 for the CD, or choose to listen to the same songs on the radio for free.

Although no one would argue that the concert is likely the best quality, not all users are able or willing to attend. They can still find value in alternative methods of accessing an artist's content. The same will be true in the Wireless Internet world. Some will be willing to pay for multimedia news clips that show full-motion video highlights of the news, sports, and weather. Others may opt for a less expensive voice and still-image version or a free simple text version.

M-COMMERCE—SECURITY PAYS OFF

Providing secure payments and protecting the privacy of an individual's personal data is critical to the growth of mobile e-commerce, also known as m-commerce. M-commerce may someday provide a true alternative to cash and make purchasing items as easy as hitting a few keys on the keypad of your Wireless Internet device.

One of the most important issues to overcome in the area of m-commerce is fraud. It's easy enough to prove payment in the physical world—cash works well and credit cards have sophisticated protection schemes to protect both the holder and the merchant from unauthorized use. Over the Internet, it's quite a bit harder to confirm the identity of the entity that is attempting to complete a purchase—merchants are unable to ask for a photo ID or some other proof that the identity of the person requesting the purchase is correct.

One solution may be *biometric identification*. Biometrics involve the use of technology that can identify unique attributes of an individual, such as a fingerprint or a scan of the face, to be presented as proof of identity. The user might have a fingerprint scanner built into their Wireless Internet device. This scanned image would be transmitted to the merchant, who would check it against a secure database of fingerprints. If the user has registered in this database, and the information matches, the merchant would allow the purchase to proceed.

SECURITY AND PRIVACY

Technology and the evolution of communications networks have been phenomenal—truly the kind of advancement that can leave users gasping in astonishment. Unfortunately, not all of the amazement is positive. Users are increasingly shocked and concerned about the lack of privacy and security in our increasingly digital world.

Whereas the majority of users are essentially unconcerned about privacy, a growing number of consumers and pro-privacy organizations are pushing for protection. In most cases it's individuals who want privacy and companies, governments, and other large organizations that want to violate it—even though this is typically done in the name of marketing, law enforcement, or even customer service. Most consumers are unconcerned because they feel they have nothing to hide—as long as there is nobody walking in the front door with a camera crew people feel they have nothing much to worry about.

PRIVACY VERSUS ANONYMITY.   Privacy has historically been something that a consumer could control to a great degree by simply choosing to whom to reveal information. Even the act of leaving your home to go shopping presented a situation where you could control who you greeted and gave your name or other personal information to. Even the time of day you chose to go shopping impacted who you might run into—going to the grocery store late at night limits the type of people who would be likely to see you. I once knew a teacher who would go shopping late at night so that she could avoid having any of her students observe her buying beer and cigarettes: She protected her privacy by controlling her actions.

This attempt at privacy didn't render my friend anonymous; privacy and anonymity are often confused because their purposes overlap considerably. Dictionary.com defines them as follows:

· *Privacy.* **1a.** The quality or condition of being secluded from the presence or view of others. **1b.** The state of being free from unsanctioned intrusion: a person's right to privacy. **2.** The state of being concealed; secrecy.
· *Anonymity.* The quality or state of being unknown or unacknowledged. One that is unknown or unacknowledged.

Anonymity can be used to reinforce privacy, however both are actually very difficult to achieve in today's society. Even those of us that are not movie stars or well-known celebrities place a high value on our privacy and ability to occasionally be unknown to those around us.

DATA COLLECTION

Often the first stage of an intrusion of privacy is data collection. If the information doesn't exist it's hard to violate one's privacy. Whereas most individuals wouldn't bother to go out and gather this information, companies do this all the time, often with our help. Grocery store loyalty programs are a good example—we willingly sell our name and other personal info along with a complete record of what we purchase and when. There are many automated forms of data collection today—frequent flyer cards, toll tags, credit cards, and phone bills. The Wireless Internet will increase the amount of data that can be collected. Initially this data will not be very personal—stock quotes, news, weather—but as devices and networks evolve to enable more sophisticated applications and services, security will become an important concern and impact the growth of the Wireless Internet.

DATA STORAGE

The next stage in the possible invasion of privacy is data storage. Data storage is becoming cheaper and cheaper. This is a great thing when consumers want to store personal content like digital pictures, tax records, and other accounting info. The bad news is that data storage has become so cheap that there is little incentive to throw data away. The majority of these systems are built to collect and aggregate data automatically, without much emphasis on error checking or data correction.

Data, both correct and incorrect, can now live on forever—errors in credit reports, medical histories, purchases, and travel records can all be accessed long after you have forgotten or even knew of them. Trails and histories of what actions took place and what content was accessed or requested become almost permanent. The digital world is unlike the physical world in that it doesn't forget the past until told to erase it. In the physical world one can be reasonably assured that eventually nobody will remember you ever visited that "unique" store or alternative nightclub back in college. Now, however, there is a digital trail back to those questionable Web sites that lives practically forever. (Or until someone deletes it, which could take even longer.)

DATA ANALYSIS AND PROFILING

Software is getting better and better at slicing and dicing and turning data into information. Databases can help companies build profiles of user behavior based on billions of statistically analyzed data points. *Data fusion* is also a popular way of adding value to data by combining two or more data sources. Your shopping history could be compared with your medical records to reveal a statistical correlation between your rising weight and your purchases of ice cream and snack foods. Processing the records of others with similar histories could result in a profile that companies could use against you. Imagine, while in line at the checkout counter, you happen to get a wireless email from your health insurance provider threatening to cancel your coverage unless you put the Ben & Jerry's back where you found it!

The very fact that Wireless Internet use will create another source of consumer data is troubling when we look at the money that is going into creating this system. The cost of new networks and services can be at least partially alleviated with revenue gained from selling this personalized data to companies that are experts in data fusion and profiling and that specialize in processing and selling consumer information.

Wonder why more consumers haven't revolted and refused to participate or even understand what is happening to their data? We've been bought off with the convenience and savings that these loyalty programs and electronic systems offer. Until consumer awareness of the potential dangers increases most will likely continue to sign away bits and pieces of personal data. We believe it will take a series of high profile abuses of personal data before many consumers will trade off that grocery store loyalty card 20 percent discount on soda and chips in exchange for greater privacy.

PERSONALIZATION GOES BOTH WAYS

Personal computers that access the Web open the door to intrusion, but the Wireless Internet will likely produce more valuable data because most devices can be tied to a person and not just a household or fixed work location. The fact that content destined for a wireless device is most often altered and filtered to conform to smaller screens and limited navigation provides even more specific data than PC surfing would generate. A PC data trail may only show a visit to a Web directory page that contains listings for entertainment; the wireless device would likely go a level deeper and reveal that a user was looking at listings for gambling entertainment. Location-based services will also add another layer of very valuable information—the history of exactly where you have been for how long.

Because technology and the data generated can be used for legitimate purposes as well as abused, we will likely not see this process of collection, storage, and analysis disappear. It may, however, eventually come under the guidance of laws and regulation that limit the potential for abuse.

Solving consumer privacy and security issues is key to enabling growth of the Wireless Internet as applications improve and become more personal. The current content accessed most commonly via wireless devices is not very personal—stock quotes, weather, general news, and the like. But future applications will enable transactions and inquiries into personal records like bank accounts and medical records—data that is personal and damaging in the wrong hands.

Surprisingly, the U.S. constitution does not currently guarantee privacy, unlike in Europe where the E.U. constitution actually guarantees a level of privacy. Many other countries have no similar laws, and in an online world where boundaries blur, even the existing laws can be hard to enforce.

FREEDOM OF EXPRESSION

The right to privacy online is linked essentially to one's ability to control disclosure of personal identity. This ability to control access to your identity is easy in the physical world: We not only decide who we interact with but our personal details are not often in danger of exposure during simple activities such as walking around town. We normally do not even provide our names unless requested (unless at a conference where no one seems to mind wearing name tags with our personal and company info for all to see). In the Internet the opposite is true— almost anyone willing to invest a little bit of effort can easily uncover the digital footprints left on any site you've visited at any date in the past.

This ability to control and hide one's identity is critical in maintaining a society that is capable of protecting freedom of expression. When government or other organizations decide that simply visiting and viewing certain information sites is a threat, the possibility of censorship takes away one of the most important powers of the Internet—freedom of expression. Individuals with differing political, religious, or lifestyle beliefs can use the power of the Internet to protect and educate others without the fear of censorship or punishment only if privacy is allowed. The transparency of the Web can be unforgiving, but never a complete picture of the whole story. Historical data that shows mistakes in an individual's past may not include enough of the data needed to paint a true picture. (PS: Will someone please tell the insurance company that the Ben and Jerry's Ice Cream wasn't for me and that my increased weight is really in error. They recorded my weight in kg instead of pounds; besides, I really have been hitting the gym, I just haven't swiped my card every time, OK?)

PROTECTING CONTENT

Not all the security focus is on protecting the user of the Wireless Internet; plenty of activity is aimed at protecting the content that users are accessing. Much of the content initially available over the Wireless Internet has been that which is available on the fixed Internet—free but not always of great value.

Studies show that Wireless Internet users are not eager to pay for the same content they can get for free on a PC. A large opportunity does exist to deliver content that users would be willing to pay for, but methods of securing this content are needed. In a physical distribution world the methods of protection are clear—pay for it and they let you leave the store with it. What you do with it after that is typically up to you (within reason).

DIGITAL RIGHTS MANAGEMENT

Digital Rights Management (DRM) focuses on methods of protecting content from theft and unauthorized distribution. DRM gives digital content publishers the ability to securely distribute high-value content such as music, books, photos, and videos in a manner that controls access and distribution. This control is central to protecting the creator's and publisher's ability to collect payments for their work.

The public has not only become used to the idea that anything found on the Internet is free but many have also shown that they will disregard copyright protection if it is convenient to do so. Software piracy has been an issue since the advent of the personal computer. Unauthorized music distribution via Napster and similar programs has been one of the fastest growing activities on the Web.

DRM is a system that controls and restricts access to the content. Authorized users may be identified individually, as a group, or even by the device used for access.

Content must first be *encrypted* or encoded to block unauthorized access. When a file is downloaded or accessed, the DRM software performs an identity check and decides if access has been paid for and authorized. If the rule for payment has been satisfied the software then "unlocks" or unencrypts the file. The file may then be accessed and used within certain parameters based on the arrangements made at time of payment. User access rights may vary just as they do in the physical world—video tapes may be rented and available for limited time periods or purchased for viewing at will.

Some DRM systems lock content to a particular access device and prevent the file from being copied or moved without authorization. Each device must have a unique identity code or serial number that is unchangeable and stays with the device for its useful life.

Files can also be watermarked and digitally encoded with information that identifies the authorized user and a record of not only when the file was transferred but what rights were granted to the user. Files that are illegally copied can be traced back to the source that was given original access.

Some of the challenges that DRM must overcome include:

- *Ease of use.* DRM will require some type of client software on the access device, and consumers may not be willing to adopt any software that limits the use of content too severely.
- *Persistent protection.* Limitations may be placed on the length of time or number of times a file is authorized for use. In other words—the music you purchase digitally today may have an authorization that expires after several years or a set number of plays.
- *Device and other sharing limitations.* Consumers are used to the right to give away or sell items that they have previously

acquired. DRM makes this difficult. *Fair-use laws* exist that allow the purchaser to make copies of music or videos for personal use (i.e., a consumer can purchase a CD and legally make a copy to play in a Walkman tape player). DRM could limit a consumer's fair use of the content.

Consumers may insist on keeping the right to transfer ownership to others in much the same way as they would other licensed content. Books and CDs are examples of licensed content that an individual can give away or sell after they are no longer wanted. DRM may impair those ownership rights.

While DRM presents us with certain challenges it will also create new and useful benefits.

- *Sticky availability or hope for the cluttered.* We all have those friends who despite help and many items from the organizer store have a unique ability to lose or "misplace" almost anything of value. DRM would allow users to simply provide a username and password (or perhaps fingerprint for those with poor memory) and gain access to all the content they have rights to even if the playback devices are all "hiding" somewhere in the closet or under the bed. Imagine how much peace and harmony could exist between teenage siblings that no longer have to argue that the other borrowed and lost their favorite CD/MP3/Movie/Game cartridge etc. A Staggering thought indeed.
- *Super distribution.* One of the more interesting benefits of DRM will be the possibility of super distribution. Super distribution is the ability to transfer content from person to person in a digital format while accounting for payments back to the publisher.

An example would be a user who has paid for and downloaded a music file; another person wanting that file could transfer payment information and authorization back to the network DRM system and receive an authorized password allowing the receipt of the file from the friend's device. The person who originally downloaded the file may be given a small commission or credit in exchange for assisting the authorized distribution of the file. In this system, it becomes possible for a service provider to collect payment not for delivering the content but for simply issuing authorization.

A FINAL WORD

As you can see there are many issues that will impact the adoption of the Wireless Internet regardless of which protocol is used or how cool the devices look. Some of these issues, such as security and privacy, will be partially sorted out in the PC world but mobility will add a layer of complexity that the Wired Internet industry has not yet begun to tackle.

The dot.com era has certainly reminded us that no matter how cool and interesting the technology, the talk will eventually turn to how to make it profitable. Content that has value must be protected from unauthorized use and copying. Mobility will challenge how pricing can be structured, bills presented, and revenue gathered in ways that do not inhibit trial and adoption.

COMMUNICATION PAST AND FUTURE

The future is about communication, but communications has always been the future!

No one would argue that the Internet has forever changed our lives. We are still inventing new ways to communicate over the Internet, ways that will continue to change how we live and do business.

If there is one thing we can count on in the future it is the need and desire for humans to communicate and share information. Throughout history we have seen countless examples of how far people will go to record and communicate thoughts, ideas, and useful information. One of the most obvious desires of mankind has been to communicate with as much detail, efficiency, and emotion as possible.

Cave painting imagery could tell an entire story regardless of the viewer's spoken language. Text-based books have been instrumental in the transfer and preservation of knowledge and understanding across cultures and continents. Music has evolved from simple human-created rhythms to complex electronic works that communicate emotions and set moods in ways that words and pictures cannot quite express.

Comparing the communication needs of our prehistoric cave-dwelling relatives to the needs of modern-day humans in skyscrapers and condos we see that some of the basics are unchanged. In terms of human communications, face-to-face interaction has not always been convenient or practical: The need to transfer thoughts and ideas *in absentia* continues to this day.

When the average consumer thinks about how they used the telephone 10 years ago and how they use wired and wireless phones today it's apparent that not much has changed. A closer look reveals something more—the ability to call anyone from anywhere at a reasonable cost has had an impact on the way society uses and reacts to technology as a whole.

As technology barrels along it often seems as if the impact is one way—that technology changes the lives of those who use it. That society is defenseless to fend off the impact of innovations that seemingly to reach into every corner of our lives. It's more accurate to think of it as a two-way street—people and culture *do* influence the future of technology albeit at differing speeds.

It's most apparent when traveling in different countries. Some cultures are very work-oriented, some value leisure and play more than material goods. The differences among cultures create the environment for technology to add value or not. How useful do you think a voicemail system or the ability to check stock quotes would be on a small island where the primary source of sustenance is fishing? The culture of societies around the world will always evolve and adapt to the opportunities before them, including technical opportunities. This is not to say that technology will never get out of hand or even cause great harm—history is full of examples of disasters caused by technical missteps. But as surely as technology may wreak havoc and destruction, cultures will bend the direction of future technology to adjust the path of evolution in hopes of more favorable results.

SERVICE PROVIDERS OF THE FUTURE

Service providers or carriers have existed in the past by charging for voice calls made on systems that use licensed spectrum. The cost of this spectrum is increasing, with huge amounts of money being paid for 3G spectrum to enable high-speed data services in addition to quality voice calls.

But licensed spectrum won't be enough for the service operator of the future. Value for wireless consumers will come from carriers that can provide end-to-end connectivity across both licensed and unlicensed spectrum. As wireless connectivity becomes more of a commodity, carriers will focus more on other aspects of service such as security, privacy, storage, and network intelligence features.

As we move towards 3G another type of service provider model will emerge, the Mobile Virtual Network Operator (MVNO). MVNOs will own no spectrum but will own or operate switches, customer care and billing systems that connect into another provider's radio system.

MVNOs—Splitting the System into Transport and Marketing

Wireless service providers have traditionally owned and operated the entire wireless system. This system included elements such as the radio transmission equipment, subscriber management systems, billing, and customer care and, of course, the license to use a certain amount of wireless spectrum. The service provider created a brand (not always with a *plan* for their brand, some had more of an accidental *reputation* than a strategic branding effort) and ran the whole system of buying spectrum, building a network and billing system, selling wireless handsets, signing up customers, and setting up customer care department to answer the phones and solve customer issues. Some service providers were better at various parts of this process than others.

Several factors have made this process more difficult as the industry matures.

- Spectrum has always been and still is limited. The sheer size and financial requirements needed to purchase spectrum in today's world favor the larger entities.
- Consumers that originally had two wireless carriers to choose from (if they were lucky) are now faced with upwards of 3–5 options in the larger metropolitan markets.

- Network equipment is increasing in capacity and available features but at the expense of increasing complexity and the need for various technical network specialists.
- Billing systems are evolving from simple time-based voice-only billing to both voice and data charges, circuit- and IP-based, and a growing number of constantly changing promotions and affinity programs not always based on time or even volume of traffic.
- Branding is becoming more important as service providers reach out beyond the traditional customer base to appeal to new groups of users each with specific needs and desired features.
- Subscriber acquisition costs average $300 and can go higher as markets reach saturation. Subsidizing handsets purchases for new subscribers ties up capital and lengthens the time to profitability for each new subscriber.
- Data capability is bringing the need for readily available content tailored to individual needs and desires.
- The costs involved in keeping the network up to date in functionality and services depends on increasing network utilization and efficiency. Even small variances in capacity utilization can be the difference between profit and loss.

So essentially the environment for the wireless service providers is one of increasing capital expense for network equipment that handles voice and data as well as provision wireless Internet content and services. The ability to leverage well-known brands to attract specific customer segments is necessary to quickly build traffic and improve time to revenue while maximizing network capacity. Service providers are looking at MVNOs as one way to increase the number of subscribers in an attempt to pay down these investments in network build out and maintenance.

Existing service providers may find that other companies with established brands in noncommunications markets are interested in leveraging their consumer relationships, content access, and knowledge by entering the mobile space as an MVNO.

MVNOs would not have the licensed spectrum needed to operate their own network, but would have other elements required offering services to the consumers. Some of the pieces an MVNO may choose to control and own include noncore network elements such as:

- Voicemail systems
- Billing systems—prepaid or subscription
- Customer care centers
- WAP servers and gateways
- Retail facilities

These noncore elements would be minimal compared to the cost of spectrum and core network equipment. Despite a limited investment in equipment, connecting to more than one existing service provider could further leverage an MVNO's brand and mobile equipment resources. This would allow the MVNO to sell services on more than one radio network. An example might be an MVNO that offers services over both a GPRS network as well as a CDMA network all under the same branding.

MVNOs would rely on the network capabilities of the underlying operator and focus their efforts on marketing and promotion efforts to build a subscriber base.

The MVNO might be a partner that has a well-known brand name and also access to content that could be resold to the service provider. Content and services could then be marketed not only under the brand of the MVNO but repackaged for the existing operators offerings.

The effect of MVNOs on existing network operators includes:

- Ability to increase the total number of users on their network albeit under two separate brands. The alternative branding offered by the MVNO could very well help an operator to appeal to new target markets not reached by the existing operators market positioning.

- Access to content and promotions made possible by the MVNO's pre-existing business. An example of could be a record label that has access to popular new music content leveraging it's artists content and brand names to attract new mobile users to their particular MVNO service.
- Faster loading of customers on newer feature rich networks could help in generating the traffic needed to maximize return on infrastructure investments.

Not all companies will have a strong enough brand to carry over into a new mobile service, some are more experienced at brand extension than others. One of the more successful MVNOs is Virgin Mobile in the United Kingdom. Launched in 1999, Virgin Mobile is a combination of Richard Branson's Virgin group (famous for Virgin record stores as well as Virgin airlines) and U.K.-based network operator One2One.

MVNOs are poised to assist network operators that wish to leverage investments in infrastructure by reselling network capacity to well know brands that are capable of providing targeted services to customers attracted to a familiar name and brand image.

Freedom from the challenges of owning and maintaining the radio transmission equipment allows MVNOs to focus on creating new services that leverage new network capabilities for voice and date while attracting and keeping customers.

Companies with strong brands could create an MVNO without the need to purchase spectrum. MVNOs would only need to negotiate with traditional carriers that have excess capacity to sell. Freedom from the challenges of owning and maintaining the radio transmission equipment allows MVNOs to focus on creating and maintaining services that attract and keep customers.

SERVICES COMPLEMENT VOICE

With all the talk about Wireless Internet applications and other data-related services it's easy to forget about voice services. Voice services will still be the bread-and-butter for service providers. Wireless Internet applications and services will largely complement voice service.

Voice calls in the future will feature new tricks such as:

- *Personal chat room effects.* The ability to allow multiple users to join and leave a conversation at will. Similar to text-based chat, users will have the ability to silence one or more users and only listen to selected callers. Just as chat rooms allow users to send private messages to selected users without leaving the main conversation, future voice calling will offer the same type of feature.

- *Stereo voice.* When voice is transmitted in stereo, users will be able to discern the direction of sounds and more readily understand who is speaking or making comments during a conference call.

- *Concierge services.* During a voice call with a concierge service representative, the user should be able to request packet data information from the live person he is talking to. For example, the caller might ask for a list of restaurants in the area he is visiting, or a map with the directions to a specific location. He can receive this document while still on the voice call and be able to discuss and clarify directions and recommendations.

- *Interactive call waiting.* During a voice call the user will be able to send a message to incoming calls to inform callers how long the wait will be or to tell them what forms of communication are currently being accepted. A notification that the user is willing to accept text messages while on a voice call would allow the caller to switch into text mode and continue the communication via text until the existing voice call is terminated.

The mixing of voice and data will allow features and services that truly fit the user's needs and preferences about convenience, detail, and environment.

4G SYSTEMS—STAY TUNED FOR 3D!

Already in the planning stages are 4G systems that allow for even more amazing voice and data possibilities. Although we don't expect to see any real 4G systems for another 5 to 7 years, working conferences on the topic were held in 2000. 4G systems based on Orthogonal Frequency Division Multiplexing (OFEM) technology are rumored to be 50 times faster than 3G, with bandwidth reaching 100 Mbps.

This capacity will enable multimedia applications such as three-dimensional (3D) renderings and other virtual experiences. Sophisticated knowledge management systems, speech recognition, and GPS also will be offered. This all implies that future wireless devices will have far greater storage and processing capabilities than current devices without increased power consumption.

CONTEXT-SENSITIVE AND USER-AWARE

When you think about PC-based Internet access there are really only two primary environments where most WWW access takes place—work or home. Not only does the majority of access take place in these two locations, the devices we use are also at least partially configured to provide easy access to the content we typically want while in these locations.

Bookmarks on my work PC are largely links to work-related topics—competitors, industry information, conference web sites, and, of course, my daily dose of Dilbert.com. The links kept on my home PC are just the opposite—music sites, news and weather sites, links to friends' home pages, and, of course, links to financial sites that do a great job of tracking my stock market losses.

The point is that PCs are already somewhat context sensitive based on the way we configure them. Mobile devices will be used in a much wider range of environments and will need to change and adapt to a user's changing needs and current context, ideally without much input or effort from the user.

Wireless Internet access devices of the future will be able to change and react to the user's environment or possible context. Data such as time of day, day of the week, location, and even events listed in network calendars can trigger the device to edit or change the presentation of information. An example is the navigation menu on a PDA that changes to reflect the user's location as being at or near work and offering icons for applications that are work related. The PDA menu would again rearrange itself when the device realized (via GPS or other location technology) that the user was now at home or in a shopping mall.

The next big leap in user-aware systems will be in the area of wearable devices and systems. Wearable computing systems might seem a bit far-fetched but humans have a history of wearing technology—eyeglasses and watches are examples of technologies that humans have become very used to wearing.

Back when clocks were primarily devices that consisted of a large wooden box with springs and pulleys, the idea of wearing one must have seemed absurd. The same is likely to occur with computers. Even now we think of computers as being at least the size of a laptop and therefore not very wearable, and certainly not very fashionable.

Many companies are advancing the science of wearable computers using Wireless Internet access built in to various pieces of clothing. One example of a company that is leading the way in wearable technology is Sensatex in Dallas, Texas. The following is an excerpt of a white paper published by Sensatex on the possibilities of a wearable computing system. (Used with permission, of course).

> Sensatex is focused on the development of the Smart Shirt System (Figure 6.1), a wearable solution for moving a wide range of information *on* and *off* an active person at *anytime* and *anyplace*. The Smart Shirt System incorporates advances in textile engineering, wearable computing, and wireless data transfer to permit the convenient collection, transmission, and analysis of personal data. By serving as an enabler of wearable computing, Sensatex expects to play a key role in the evolution of personalized, mobile information processing.

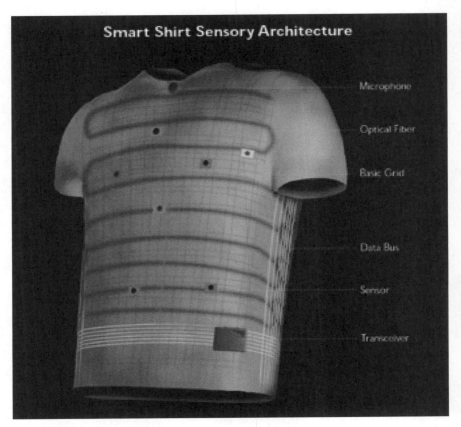

FIGURE 6-1   Smart shirt.

The Technology

The System incorporates the Wearable Motherboard™ Smart Shirt (the "Smart Shirt" or "garment"), a novel electro-optical garment funded by the Defense Advanced Research Projects Agency (DARPA) and developed at the Georgia Institute of Technology, and an advanced communications and data management infrastructure. Together, this integrated solution provides an extremely versatile network for sensing, monitoring, and information processing devices that can enable a wide range of products.

The Smart Shirt permits the seamless acquisition and disposition of sensory and environmental data to and from a wearer; and the communications and data management infrastructure permits the transmission of this data over wired and wireless networks. The Smart Shirt's

Interconnection Technology provides a flexible "bus" structure that allows a potentially enormous array of sensors, whether physical, environmental, or biological, as well as information processing devices, to be mounted or embedded at any location. These flexible capabilities permit data to be collected from the wearer in an *unobtrusive* fashion and routed to and from the communications and data management network. By providing hands-free bi-directional data transmission, data processing systems, and optimized wireless communications using an expedient array of protocols, Sensatex offers a total solution that is virtually transparent, but ubiquitously available to the end user.

The Products
By offering a customizable range of information processing capabilities under mobile settings, wearable computing technology finds applicability across the consumer reassurance, lifestyle enhancement, and healthcare monitoring markets. A wide breadth of the population, from infants to the elderly, can use devices such as the Smart Shirt System to enjoy meaningful improvements in the quality of life.

REASSURANCE PRODUCTS

Infant/Toddler/Active Child Monitor

The Sensatex Child Monitor is a "digital umbilical cord" that allows parents to keep track of active children. The initial products in this category will offer parents two forms of functionality—continuous two-way voice communications with children in the home environment and/or ubiquitous communications with children, regardless of the parents' location. The next generation product of the Sensatex Child Monitor, whether local or wide area, will provide the added features of vital signs monitoring and GPS-based locator services.

Geriatric Monitor

The Sensatex Geriatric Monitor provides reassurance to seniors and their families by continuously monitoring vital signs, being able to detect a fall, knowing location through the use of a GPS chip, communicating with two-way voice, offering a "panic" button, and connecting automatically with emergency services via pre-configured parameters.

CONSUMER PRODUCTS

Amateur/Individual Sports

Triathletes, marathon runners, cross-country skiers, cyclists, mountain climbers, rowers, and aerobic sportsmen, among others, present a significant market for the Sensatex Smart Shirt. The Smart Shirt System will monitor heart rate, respiration rate, body temperature, external temperature, altitude, location, and orientation (compass) for the amateur athlete, with the added benefit of hands-free connectivity and switching capability between an MP3 player, radio, cellular phone, and voice recorder. In addition, the product will provide voice-activated readout for hands-free access to information (e.g., a runner queries "distance" via voice and receives voice response of "12.6 miles").

Team Athletic Training

The team model will utilize the basic architecture of the amateur athletic training product in a kit that also includes computing, communications, analytical software tools, real-time display, and Internet access, allowing coaches and trainers to manage the training of athletes in a highly efficient manner.

Personal Area Networking

As a powerful wearable solution, the Smart Shirt System can act as an enabler of the personal area network (PAN). As computing devices become smaller and more portable, people have taken to carrying notebook computers, pagers, PDAs, and cell phones nearly everywhere they go. The next technological step is to connect these disparate devices and create a single network that would relieve the user from the redundancy of features and concern over where information is stored. As third generation cellular networks are introduced and wireless local area networks and Bluetooth technology proliferate, Sensatex will be able to facilitate the deployment of the PAN using the Interconnection Technology of its Smart Shirt and the unique features of its communications and data management infrastructure. The System's unique advantage is its ability to interface with devices both wirelessly and with wires, thus permitting a total solution to generate optimal connectivity.

MEDICAL PRODUCTS

Infant Vital Signs Monitoring

The Smart Shirt System will support the *unobtrusive* collection of vital signs data (EKG, respiration rate) for the detection of various disorders in infants, such as apnea, prematurity, respiratory synctial virus, gastroesophageal reflux, and seizure disorders. The data collected off the infants' body is transmitted wirelessly to a processing location that will utilize analysis software provided by an existing infant monitor manufacturer.

Sleep Studies

Despite a vast number of U.S. adults with sleep disorders, such as adult apnea, insomnia, and narcolepsy, only a small fraction seek diagnosis due largely to the discomfort and inconvenience of hospital-based testing. The Smart Shirt System offers test subjects a comfortable and unobtrusive means by which to have their vital signs monitored and transmitted for the diagnosis of these disorders.

Hospital/Nursing Home Monitoring

Sensatex will partner with patient monitor manufacturers to develop a Smart Shirt System that transmits vital signs data either with or without wire to the vendor's stationary or portable monitoring unit(s). This includes retrofitting monitors already manufactured and sold by the vendor to hospitals, nursing homes, or skilled nursing facilities. By acting as a single platform for all sensory components, the Shirt will facilitate the placement of sensors on the patient and will either entirely eliminate or greatly reduce the number of wires connecting the patient to the monitor. The sensor suite for this product will include 3-lead EKG, pulse oximetry and respiration rate.

Home Monitoring

Sensatex is actively investigating the home-based health monitoring markets for conditions that warrant continuous and/or near real-time vital signs data collection and transmission. An effective remote health monitoring product could play a major role in the industry-wide trend to transport patients to lower acuity and lower cost settings without sacrificing the quality of care. At home, or even while being transported via ambulance, the Smart Shirt System will be able to collect health data in a comfortable and unobtrusive manner and transmit it to a facility (including the emergency room) that possesses optimum expertise about the patient's condition. The cardiac patient population, including those prescribed Holter and event monitors, could significantly enhance the quality of their monitoring using the Smart Shirt System technology.

MILITARY AND OCCUPATIONAL PRODUCTS

Battlefield Combat Care Solution

The Smart Shirt was originally funded by DARPA for the purpose of detecting bullet penetration, transmitting soldiers' vital signs to a remote medical triage unit, and ultimately saving lives on the battlefield. Sensatex expects to work with DARPA and the Department of Defense to complete the design of an enhanced Smart Shirt System for the 21st Century Land Warrior program. This System will be able to fully withstand battlefield conditions and interface with the military's existing radio systems to create full Information Node Capability for the individual soldier.

Hazardous Occupations Monitoring

The Smart Shirt System has the potential to play a vital role in the health and safety monitoring of individuals working in hazardous conditions. Monitoring devices could be developed that incorporate vital signs, physical sensors like GPS, and targeted environmental sensors, such as carbon monoxide, other poisonous gases, and temperature.

As we can see from the activities of even one company that is focused on wearable technology, this is an area that has incredible potential as Wireless Internet continues to evolve. Although many future devices may be in fact wearable, not all future wearable devices will have full computing capabilities. Even now, basic cellular phones are worn on users' belts, in much the same way as pagers. It may be a long time before wearable computers and other devices are as common as other wearable technology but don't be surprised if the youth of the future trade in their Sony Walkman and MP3 players for computer-capable clothing. (Perhaps this means that The Gap and Radio Shack will someday have something in common, too.)

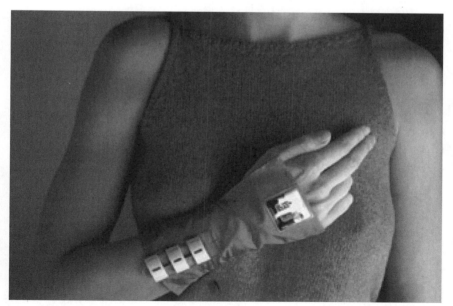

FIGURE 6-2   Concept idea—wearable phone. Courtesy of Motorola, Inc.

ECONOMIC POWER

The Wireless Internet raises consumer power. By using the Wireless Internet, people will become used to having as much information as they need when making a significant buying decision. Their information searching and digestion no longer needs to be done at home: They will be able to access more information outside the home, virtually anywhere, and at the right time. For example, when buying a car, a user could have all the latest information, including price quotes from competing dealers, when visiting a dealer for a test drive.

WLANS AND BLUETOOTH— THE NEW ACCESS POINTS

While Internet access at work or home is easily accomplished with fixed connections, wireless access is beginning to come up to speed and offer the benefit of mobility albeit over short distances. This growing wireless access technology is called Wireless Local Area Network (WLAN) access and is essentially a networking system that creates a wireless connection between a device, typically a laptop PC, and a network or the Internet. The power and size requirements of WLAN device cards make it more suitable for laptops and some PDAs than for smaller handheld devices such as cellular phones.

The majority of current WLAN systems use a technology called the 802.11 standard, also known as *Wi-Fi*. Wi-Fi began a couple of years ago under the name IEEE 802.11 High Rate (HR) standard. Someone with a sense of nostalgia coined the easier to relate term of Wi-Fi as a play on the old audio term hi-fi (high fidelity), which entered the language back in the 1950s.

One of the key features of WLANs is that they use unlicensed spectrum, currently most in the 2.4 GHz range although systems are possible in 900 MHz as well as 5 GHz. Operating in unlicensed spectrum allows WLAN operators to set up a wireless network for only the cost of equipment and its connection to a network or the Internet.

These WLANs are useful for providing access in targeted locations such as offices as an extension of the wired network or in public areas such as airports, hotels, and shopping centers. These 2.4 Ghz systems typically have a range of about 300 feet and currently offer speeds of about up to 11Mbps, which is roughly equivalent to older 10 Mbps Ethernet throughput. Future WLAN standards should increase this to up to 54 Mbps which would be much more suitable for multi-media applications such as Video.

WLAN technologies can be implemented in two ways: *access point* or *peer-to-peer configuration.* Access point configuration is the most popular method and uses a wireless point of access that connects to the fixed network connection on one end and transmits wireless signals on the other end. Access points typically accommodate six network interface cards (NICs). These PCMIA type cards are the devices that allow the connection back to the access point.

The other method of implementation is the peer-to-peer configuration that works by allowing each end client or device card to connect to each other and communicate directly between laptops or devices.

Most WLANs are currently installed in offices to provide mobility to workers that need to access the network while away from their desks a growing number of systems are being setup for public use.

There are groups of socially minded individuals that are using public access WLANs to create pockets of network access that would be free for all to use. A basic WLAN system can be installed for under $1,000 and offer service to users in a 300 foot radius of the antenna. Anyone with an 802.11b card could in theory access the network. These networks could offer limited access to localized content on a community oriented intranet as well as offering access to the broader Internet.

WLANs are also being planned for public access in locations that consumers frequent including airports, coffee shops, and hotels. The potential for these systems to offer localized content including advertisements or event specific content has businesses excited about the possibilities. Starbucks has announced plans to offer WLAN access points at numerous test locations that would give anyone with a WLAN card in their device access to a network. This network could be provided as a free service or offered as a pay as you go system. These systems can provide more than basic Internet access, Starbucks could provide information designed to improve customer service and sell more coffee.

BLUETOOTH—SHORTER RANGE BUT LOWER POWER CONSUMPTION

For devices that are too small or unable to provide the power needed to operate a WLAN card there is a new technology in the works called Bluetooth. Similar to WLAN technology, Bluetooth is designed for smaller devices with lower power consumption. This lower power means a shorter range of transmission; Blue tooth will initially be limited to approximately 30 feet. With potential throughput on par with WLAN, Bluetooth will compete with WLAN for short range access for laptop PCs and printers. Bluetooth will be more suitable for smaller devices and operates on a peer-to-peer configuration that establishes a *piconet* (a series of connected devices that pass the data along to one another in a kind of fire brigade manner) to extend the reach back to a network when available. Bluetooth will be used more often in scenarios where data is to be exchanged between two or more devices such as between multiple PDAs or between a PDA, cell phone, and a printer.

CELLULAR THREAT OR BENEFIT?

Some may see WLAN and Bluetooth access as competition for cellular access, it's more likely to evolve as an extension of cellular operators offerings. Wireless operators need WLAN technology to offload heavy localized traffic especially in indoor environments. Recent studies support the claim that 3G operators will need WLAN technology to handle the most demanding users in dense areas. The fact that WLAN uses unlicensed spectrum will not mean that operators wouldn't need to spend money to operate these systems as they will need the same centralized network management, subscriber identification, security, and roaming systems in place to manage this last portion of the wireless transmission. WLANs could be configured to cache frequently accessed content and reduce the data traffic carried by the backbone of the system that would be using licensed spectrum.

Other issues ill need to be overcome for WLAN technology to peacefully exist with other systems that use the unlicensed spectrum. Congestion is a real possibility as there is no limit to the number of networks and traffic that could try to use the frequency in a given area. Security is an issue that is growing in importance as more and more users have a WLAN card and can easily walk or drive within range of many corporate systems. Many of these systems have been installed with no change to the equipments factory settings enabling anyone that knows the common factory settings and passwords to gain access to corporate networks. Interference will also be a challenge especially for companies such as Starbucks since other devices such as Microwave ovens can generate frequencies close enough to 2.4 GHz to disrupt the WLAN transmission every time they heat up a muffin.

CHANGING THE ROLE OF WIRELESS OPERATORS

Wireless technology is still in its infancy and as technology changes so does the business model of the service providers. Rarely have major changes to the infrastructure failed to change the possible business models.

As technologies such as WLAN and Bluetooth improve, they will serve a portion of consumers need for connectivity. Wireless operators have traditionally offered services only on licensed spectrum as it protected them from undue competition. The future will see operators offering services over both licensed and unlicensed spectrum reflecting a shift towards the need for seamless management of wireless access across various networks.

Operators of the future will need to compete not only against the few that have purchased licensed spectrum but also those that offer services in the unlicensed band. We believe that most enterprise and consumer users will lean towards solutions that integrate the authentication, customer care, and billing of as many network access technologies as needed for their particular use. Operators with licensed spectrum will still have an edge over those without licenses, as many users will resist the hassle of separate subscriptions for access needed on different overlapping networks that will often service the same laptop or mobile device.

DIGITAL DIVIDE—HOW WIRELESS CAN CHANGE THE WORLD

The growing consensus is that in the New Economy access to knowledge is critical for economic success. Unfortunately the economic power of the Internet is not equally distributed.

Recent Internet usage statistics show that there are currently 429 million Internet users worldwide. This number is actually small when considered in context. Of that 429 million, 41 percent are in North America; in fact, the United States has more computers than the rest of the world combined!

These 429 million users actually represent only 6 percent of the world's entire population. The following breakdown shows just how uneven Internet usage is across the world's regions.

Of the online population:*

· 41 percent are in the United States and Canada
· 27 percent live in Europe, the Middle East, and Africa
· 20 percent are located in Asia
· Only 4 percent are located in South America

The importance of Internet access will further divide the world's population into two main groups—those having access and those who do not.

* Source: First Quarter 2001 Global Internet Trends, Neilson/Netratings.

The poorest members of society suffer based on three primary assumptions:

- The poor cannot afford to buy the necessary equipment needed to be connected to the Internet.
- The infrastructure of developing countries may be so poor that a significant portion of the population is not able to connect even if equipment is available.
- The poor may not be literate enough to make use of equipment and connectivity even when available.

The issue of the digital divide is beginning to evolve into a drive towards realizing the *digital dividend*. The digital dividend focuses on how to use technology to improve the economic possibilities of global society.

Some of the key principles that will enable a digital dividend include:

- *Access vs. ownership.* The assumption that users must purchase equipment to have access to the Internet must be challenged. In the New Economy the true economic benefit comes from access to sources of knowledge and competence, not from ownership of the access device.

  A phenomenon is developing in several developing countries where the trend is for individuals with equipment and access to create a business around providing access. Local entrepreneurs in India (mostly women) are operating pay-per-use telephone services that provide traveling access to remote and other underserved areas. With little more than a mobile phone, these entrepreneurs have made access to the telephone possible for a large number of urban poor and people in remote villages. Many are now adding fax and PC services to their portfolio of services.

- *Rational trade offs.* While many of us would opt for direct ownership of a PC or cellular phone, trading currency for convenience, the poor make an equally logical trade-off by exchanging personal convenience for low-cost, no-investment access.

This approach may also make sense for those who are able to purchase, because technology seems to advance at a rate that quickly makes equipment obsolete!

In an age of ever-changing PC features, individual ownership may not be the best choice after all.

- *The connectivity leapfrog.* Many developing countries have never had far-reaching telephony systems due in part to the cost of infrastructure needed to cover sparse or difficult terrain. Without a legacy wireline system in place, users are unable to access even simple communications. With infrastructure costs less than half that of a wireline system, wireless is becoming the telephony system of choice for many regions that lack existing copper connections to homes and businesses. The Wireless Internet will help overcome connectivity issues in countries that lack adequate physical wiring.

- *Multimedia literacy.* It's well known that the Internet started as largely an English-language medium to the exclusion of many languages, especially those that use a non-Arabic alphabet. The tide is slowly turning and more Web sites are publishing content in local languages.

  The move towards multimedia will also help alleviate this issue for those who are not able to read text but can communicate verbally and visually. Many cultures have unique dialects that are difficult and costly to translate into text but that can be published at lower cost in a voice format.

  Multimedia will enable communication to take place in ways that accommodate the needs of the user by integrating text, audio, and video in ways that the individual user can utilize.

WIRELESS BRIDGES THE DIVIDE

A Wireless Internet can play an important role in transforming the digital divide into the digital dividend. The flexibility of wireless infrastructure allows carriers to provide coverage in difficult terrain as well as access in established buildings with minimal labor and installation time. Equipment costs are much less than for the PCs typically used to access the fixed

Internet; therefore the Wireless Internet is more accessible for those who wish to own the access equipment for personal use as well as for pay-per-use businesses. Remote users in developing countries will benefit from the mobility and freedom of smaller more portable devices that can be easily transported from village to village. Although Wireless Internet access is more limited than fixed, PC-based access, many countries will benefit from the use of wireless access services as an important part of the digital dividend solution.

WIRELESS INTERNET— THIS TIME IT'S PERSONAL!

Our personal information is increasingly found in digital format—pictures, letters, bills, receipts, videos—digital means it's easier to share not only content but the impact of content, whether the content is informational, educational, entertaining, or emotional.

Internet users today can create messages that incorporate many media types: emails can include attachments of sound, picture, audio, and pure data files. But let's think about *what* we would send and *when* we send it once we have the ability to compose and send while mobile. In short, real-time distribution will result in an increase in the quality and frequency of communication.

LIFE TURNS DIGITAL

We possess increasingly more personal digital content—digital photos and video clips, digital music clips, and even cherished emails. (Admit it—you've saved more than one personal email for no other reason that to read it over and over, you softie!) The Wireless Internet will encourage the collection of a growing amount of personal digital content. Some of the newer wireless devices have already announced plans for MP3 players, audio-recording capabilities, and built-in digital cameras.

We will soon have the tools to digitally capture and share like never before. Just as the world was forever changed with the adoption of the personal video camcorder. (As chronicled by shows like *America's Funniest Home Videos*—just imagine what the future will bring. Anyone care to tune into *America's Funniest PDA Audio Captures*? Just think: You could actually win a prize for recording those Dilbertesque comments your boss makes in the weekly staff meetings.)

TECHNOLOGY IMPROVES SOCIAL INTERACTION

It is easy to think that all this new technology will dehumanize us all and shift the emphasis from communicating with people to interacting with technology. But the reality is just the opposite, because innovations such as the Wireless Internet allow for more frequent and detailed social contacts.

The Wireless Internet will be used more as a social medium, making complex interactions less dependent on face-to-face encounters. Technologies such as wireless email and messaging help maintain contact while away from friends and family and are very useful for arranging impromptu face-to-face interactions. There will be an evolution from using a voice-only phone to using a 2.5G or 3G computing or handheld device to send pictures, coordinate diaries, organize social events, and play games.

People will also benefit from the multimedia presentation of information. The inclusion of graphics, sound, and animation as part of the information that users consume conveys much more than text. In an age of macromedia Flash and MTV today's users may reject information that is not presented in an interesting way.

MULTIMEDIA MESSAGING

One of the Wireless Internet technologies on the near horizon is Multimedia Messaging (MMS). MMS is an application that uses a data call to a wireless device that delivers a message capable of incorporating any of the following in an organized and choreographed presentation:

- Pictures
- Data
- Text
- Audio
- Video
- Voice

Whereas SMS messaging typically uses a digital control channel, MMS will be one of the first applications that make use of the carrier's higher speed data capabilities available in 2.5 and 3G systems. MMS will take advantage of combinations of media to allow users to communicate with more detail, emotion, and efficiency. It's important to understand how wireless mobility adds value to multimedia by allowing the *timely* exchange of information.

MMS will be used to communicate in ways that even a digital voice call can't achieve. Though many of us have tried to explain the sights and sounds around us while on a simple voice call we can agree that the effect is poor at best. Just as SMS will be the first nonvoice communications most of us encounter, MMS will be one of the first 3G communications we use in a wireless fashion.

WIRELESS EFFICIENCY

As we mentioned earlier, humans have always sought to communicate efficiently. Who wants to endlessly repeat something or have to deal with not being understood? The most successful persons throughout history have been those who communicated well on some level. Perhaps it wasn't through speech—an engineer might choose a technical drawing to entirely communicate an idea and avoid talking at all.

IT'S ALL ABOUT EMOTION

We all remember the AT&T long distance ads on television that encouraged us to "reach out and touch someone." Despite what AT&T might have charged back in the good old days of the long distance monopoly, we must admit that they had figured out the most important driver of communication. They realized that personal communication is largely an emotional activity, and people will pay to share emotions. Now, we aren't saying that communication should make you cry, but communications can allow the kind of sharing that people will value.

NO, I DON'T WANT TO SEE WHAT YOU DID LAST SUMMER

Photos are a great example of sharing emotion. Think about the pictures the average family takes: The subjects are people and places that they care about—family and friends, places they visited, etc. The activities pictured add more detail to a child's birthday party, a friend's graduation, or scuba diving on that Caribbean vacation.

Now think about what you do after you get the pictures developed (assuming you don't leave the film in a drawer for a year). The natural inclination is to show others. Why do we do it? To share the emotions that we felt when the pictures were taken. Whether you were there when the picture was taken or not, you're still fair game when those pictures come back from the photo lab.

ERODING EMOTION

One big problem exists: Emotions erode with time. You'll put up with co-workers who pull out pictures of some recent event but most of us tend to run when someone suggests sitting down to view that old home movie or pictures from that vacation back in 1978. Newlyweds always seem to have a wedding album handy but grandma and grandpa have theirs packed away someplace (if you're lucky).

So the goal becomes trying to share emotions in a timely manner—in near-real-time whenever possible. Whether it's an IM session giving you a blow-by-blow account of the heated debate coming from the corner office, a newly snapped pic of the goings on down at the local pub, or a recently recorded audio clip from your friends at the concert that you couldn't get away for, communication offers more emotional value when it is *timely and fresh*. Wireless Internet applications will help users make the most of personal communications while the content to be shared still has value, before it erodes and becomes lifeless and dull.

SPEED INFLUENCES THE VOLUME OF COMMUNICATION

The speed of our communication process influences the amount of things we want to communicate. Real-time communications allow us to share things while they still have relevance. Human communication is often about human experiences—things that somehow impact our five senses. Even intense experiences eventually fade from our memory. Communication of these experiences is best right after the event, or ideally, during the event.

When was the last time you took pictures? Birthday party? Vacation? Wedding? Pictures are usually taken at high-emotion events so that we can capture the moment and remember it later. How fun is it to share these kinds of photos with friends soon after you take them? Many of us can't wait to share our pictures as soon as we get them. Trouble is, the longer you wait, the less fun it usually is to share. As Wireless Internet devices enable users to capture and transmit images, sound, and other data the frequency of communication will increase.

REAL TIME ADDS VALUE

Remember the last time you went to a concert or show? Let's assume you have one friend who would have enjoyed the show but wasn't able to attend: The longer you wait to tell him about it the less you will remember and the less emotion *you* will feel about the event. As time passes, you'll have a reduced ability to recall the event details.

Now imagine that you could share images, sounds, and your thoughts in text in almost real time. Ever watched a live TV show? The value of sharing events while the event is occurring is apparent on TV. Wireless multimedia messages won't be TV, but they will be more like a short commercial—images, sounds, and text combined to communicate with detail, efficiency, and emotion and to allow the person on the other end to better understand you.

THE FUTURE OF WIRELESS INTERNET IS CERTAIN—TO CHANGE!

This book has covered the technologies and applications of the Wireless Internet in an attempt to give you a high-level glimpse of the many challenges and issues surrounding its evolution. On many levels it's still anyone's guess as to which protocols and specific technologies will emerge as part of the standard of the future.

The wireless industry has many players all working to provide their contribution to this amazing future of Wireless Internet access. Not everyone agrees on the best way to ensure success but the momentum has generated a self-fulfilling prophecy of sorts, led by industry: If they think it will happen, it will (eventually anyway). But how much will it cost consumers and industry? Who will profit?

WILL THE WIRELESS INTERNET SURVIVE?

In an age where rapid technology development produces concepts and innovations that disappear often as quickly as they come it's only natural to ask the question—Will the Wireless Internet survive? We believe the Wireless Internet will *eventually disappear*.

It will be out of sight, but it will still exist. Not as the wired or wireless Internet, but simply as "the Internet" or "the network." Access method and device will eventually become irrelevant.

As the Wireless Internet evolves and embeds itself in the society and culture of our modern world, the phrase "Wireless Internet" will quietly go away. When is the last time you heard someone refer to the "electric" light? Or the "gasoline powered" automobile? Or even "indoor" plumbing? The descriptors of *how* eventually fall away as society gets used to assuming the obvious or irrelevant. What will matter in the future is that a user is connecting to a network; whether that user arrives via cable broadband, GPRS, or a public WLAN won't really matter.

BRANDWIDTH OVER BANDWIDTH

Branding will eventually replace the how. Just as consumers talk about fueling up down at the local Texaco station but don't bother to explain if they filled up with diesel or gasoline, they will talk about accessing the network via a particular brand. Wireless Internet access brands will be not unlike the cellular carriers of today; I'm an AT&T customer but often roam or use another carrier's network all while telling others I'm an AT&T customer.

Even the most insightful futurists can't guarantee exactly what the interaction between culture and Wireless Internet technology will result in. But even though the experts can't predict how the Wireless Internet will evolve, please keep one thing in mind—the answer may someday be in your hand.

APPENDIX A

SUGGESTED FURTHER READING

3G Wireless Demystified, Lawrence Harte, Richard Levine and Roman Kikta, McGraw-Hill, 2001, ISBN: 0-07-136301-7.

Delivering xDSL, Lawrence Harte and Roman Kikta, McGraw-Hill, 2001, ISBN: 0-07-134837-9.

CDMA IS-95 for Cellular and PCS: Technology, Applications, and Resource Guide, Lawrence Harte, Roman Kikta, and Daniel McLaughlin, McGraw-Hill, 1999, ISBN: 0-07-027070-8.

Inside WAP Programming Applications with WML and WML Script (With CD-ROM), Pekka Niskanen, Addison-Wesley Pub Co, 2000, ISBN: 0201725916.

Beginning WAP: Wireless Markup Language & Wireless Markup Language Script, Soo Mee Foo, Ted Wugofski, Wei Meng Lee, Foo Soo Mee, Karli Watson, Wrox Press Inc, 2000, ISBN: 1861004583.

WCDMA: Towards IP Mobility and Mobile Internet, Tero Ojanpera, Ramjee Prasad, Artech House, 2001, ISBN: 1580531806.

Wireless Computing : A Manager's Guide to Wireless Networking, Ira Brodsky, John Wiley & Sons, 1997, ISBN: 0471286567.

Wireless Lan Systems (The Artech House Telecommunications Library), A. Santamaria, F.J. Lopez-Hernandez, Asuncion Santamarie, Artech House, 1994, ISBN: 0890066094.

Wireless Web: A Manager's Guide, Frank P. Coyle, Addison-Wesley Pub Co, 2001, ISBN: 0201722178.

Advanced Internet Programming, Sergei Dunaev, Charles River Media, 2001, ISBN: 1584500603.

Bluetooth Demystified (McGraw-Hill Telecom), Nathan J. Muller, McGraw-Hill Professional Publishing, 2000, ISBN: 0071363238.

Bluetooth Revealed: The Insider's Guide to an Open Specification for Global Wireless Communications, Brent A. Miller, Chatschik

Bisdikian, Anders Edlund, Prentice Hall PTR, 2000, ISBN: 0130902942.

WCDMA for UMTS: Radio Access for Third Generation Mobile Communications, Harri Holma, Antti Toskala, John Wiley & Sons, 2000, ISBN: 0471720518.

The GSM Network: GPRS Evolution: One Step Towards UMTS, Joachim Tisal, John Wiley & Sons, 2001, ISBN: 0471498165.

Cdma Mobile Radio Design (Artech House Telecommunications Library.), John B. Groe, Lawrence E. Larson, Artech House, 2000, ISBN: 1580530591.

Handbook of Radio and Wireless Technology, Stan Gibilisco, McGraw-Hill Professional Publishing, 1998, ISBN: 0070230242.

The Cell Phone Handbook : Everything You Wanted to Know About Wireless Telephony (But Didn't Know Who or What to Ask), Penelope Stetz, Aegis Pub Group, 1999, ISBN: 1890154121.

Cellular and PCS: The Big Picture (McGraw-Hill Series on Telecommunications), Lawrence Harte, Richard Levine, Steve Prokup, McGraw-Hill Professional Publishing, 1997, ISBN: 0070269440.

The Comprehensive Guide to Wireless Technology, Lawrence Harte, T. Schaffnit, Steven Kellogg, Steve Kellogg, Richard Dreher, Nancy Campbell, Lisa Gosselin, Judith Rourke-O'Briant, APDG Publishing, 2000, ISBN: 0965065847.

APPENDIX B

WIRELESS NEWS/OPINION

Unstrung
  www.unstrung.com

Red Herring "Wireless Watch"
  www.wired.com/news/wireless

Unwired News
  www.redherring.com

About Telecom
  www.telecom.about.com

Wireless Newsfactor
  www.refreq.com/industrylinks.htm

3G Newsroom.com
  www.3GNewsroom.com

WIRELESS INDUSTRY NEWS

Wireless Week
  www.wirelessweek.com

CTIA News
  www.wow-com.com/news

TelecomClick
  www.telecomclick.com

RCR
  www.rcrnews.com

Mobiledatabiz.com
  www.mobiledatabiz.com

Wireless Review
 www.wirelessreview.com

Wireless Internet Magazine
 www.wirelessinternetmagazine.com/

Broadband Wireless News
 www.shorecliffcommunications.com/magazine/index.asp

EMAIL NEWSLETTERS

Unstrung
 www.unstrung.com/index.php3

CTIA Daily News
 www.wow-com/news

WIRELESS DEVICES

AllNetDevices
 www.allnetdevices.com

PDABuzz
 www.pdabuzz.com

Thinkmobile.com
 www.thinkmobile.com

WIRELESS ADVERTISING

Wireless Advertising Association
 www.iab.net/waa

WIRELESS RESEARCH FIRMS

Herschel Shosteck
 www.shosteck.com

Strategy Analytics
  www.strategyanalytics.com

Cahners In-Stat
  www.instat.com

Yankee Group
  www.yankeegroup.com

Forrester Research
  www.forrester.com

Jupiter Wireless
  www.jxwireless.com

INDUSTRY LINKS

Refreq.com
  www.refreq.com/industrylinks.htm

Group 3G
  www.3Gportal.com

3G.IP
  www.3GIP.org

3GSM World Congress 2002
  www.3gsmworldcongress.com

Mobile Applications Initiative
  www.mobileapplicationsinitiative.com

The Road to 3G
  www.zdnet.co.uk/news/specials/2000/08/road_2_3g/

Ericsson—Introduction to 3G
  www.ericsson.com/3G/

Third Generation Partnership Project (3GPP)
  www.3gpp.org/

EdgeMatrix
  www.edgematrix.com/

CEO Mobile
  www.ceomobile.com/

eMobileNet
www.emobinet.com/

Wireless in a Nutshell
www.wirelessinanutshell.com/

Oracle Mobile
www.oraclemobile.com/

Industry Statistics
www.cellular.co.za/stats/stats-main.htm

The Telecom Corridor
www.telecomcorridor.com/tc/index.htm

Total Telecom
www.totaltele.com/

Java Mobiles
www.javamobiles.com/

wireless.internet.com
www.wireless.internet.com/

MobileGPRS
www.mobilegprs.com/

internet.com
www.internet.com/sections/wireless.html

Carriers World
www.carriersworld.com/

Corporate Wireless Group
www.goam.corporatewireless.com/wireindex.htm

TIA B2B Glossary
www.tiab2b.com/glossary/

WAP SITES

2 Thumbs WAP.com
www.2thumbswap.com

[WAP] Resource Kit
www.macromediatraining.net/wap/

Eazywap
  www.eazywap.com/

Nokia WAP Developer Forum
  www.forum.nokia.com/main/

Gelon
  www.gelon.net/

Jumbuck
  www.jumbuck.com/

The WAP Forum
  www.wapforum.org/

The WAP Trap
  www.freeprotocols.org/wapTrap/

WAP Forum—W3C Cooperation White Paper
  www.w3.org/TR/NOTE-WAP

WAP.com
  www.wap.com

WAPCardz.com
  www.wapcardz.com/

WAPinside News
  www.wapinside.com/

WAPsight
  www.wapsight.com/

Wapsody
  www.alphaworks.ibm.com/aw.nsf/techmain/

WAPuSeek
  www.wapuseek.com/

IEC WAP Tutorial
  www.iec.org/online/tutorials/wap/index.html

GSM World WAP Tutorial
  www.gsmworld.com/technology/wap.html

Speedy Tomato
  www.speedytomato.co.uk/

WebCab.de
  www.webcab.de/

WMLScript.com
www.wmlscript.com/

WOAOP.com
www.woaop.com/

SOFT SWITCHES

The SIP Center.com
www.sipcenter.com/

Softswitch Consortium
www.softswitch.org

SIP Forum
www.sipforum.org/

INDUSTRY ASSOCIATIONS

Softswitch Consortium
www.softswitch.org

Bluetooth
www.bluetooth.com

Canadian Wireless Telecommunications Association
www.cwta.ca

GSM Association
www.gsmworld.com

International Mobile Telecommunications Association
www.imta.org

Mobile and Portable Radio Research Group
www.mprg.ee.vt.edu

Mobile Data Association
www.mda-mobiledata.org

Personal Communications Industry Association
www.pcia.com

Universal Wireless Communications Consortium
www.uwcc.org

Wireless Communications Alliance
www.wca.org

The Wireless Foundation
www.wirelessfoundation.org

Broadband Wireless Internet Forum
www.bwif.org/

TIA Online
www.tiaonline.org/

Mobile Wireless Internet Forum
www.mwif.org/

Cellular Telecommunications & Internet Development
www.wirelessdata.org/front.asp

WIRELESS LANS

IEEE 802.1Q VLAN support for FreeBSD
www.euitt.upm.es/~pjlobo/fbsdvlan.html

Short Tutorial on Wireless LANs and IEEE 802.11
www.computer.org/students/looking/summer97/ieee802.htm

Wireless Ethernet Compatibility Alliance
www.wi-fi.org

802.16 Tutorial
www.ieee802.org/16/tutorial/index.html

GLOSSARY

2G *Second generation.* Generic name for second generation of digital mobile networks (such as GSM, and so on).

2.5G An interim solution between 2G and 3G.

3G *Third generation.* Generic name for next-generation mobile networks (Universal Telecommunications System [UMTS], IMT-2000; sometimes GPRS is called 3G in North America).

3GPP *3G Partnership Project.* An industry standardization partnership for W-CDMA.

3GPP2 *3G Partnership Project 2.* An industry standardization partnership for CDMA.

AAL2 *ATM adaptive layer 2.* Supports connection oriented traffic, compressed voice and data.

ADSL *Asymmetric Digital Subscriber Line.* Used for Internet access, where fast downstream is required, but slow upstream is acceptable.

AM *Amplitude Modulation.* Modulation in which the amplitude of a carrier wave is varied in accordance with some characteristic of the modulating signals.

AMPS *Advanced Mobile Phone Service.* The first generation of cellular analog service. Although there were previous analog mobile phone systems, this was the first to use a cellular structure.

ANSI *American National Standards Institute.*

API *Applications Programming Interface.* A set of functions and procedures for developing an application. The core set of facilities made available to the developer/programmer for writing applications like system functions and procedures for manipulating information.

ARPANET *Advanced Research Projects Agency Network.*

ARPU *Average Revenue Per User.*

ASP   *Application Service Provider.* A company that hosts software applications on its own servers within its own facilities.

ATM   *Asynchronous Transfer Mode.* A digital transmission system using 53-byte packets. ATM may be used for LANs and WANs.

Bandwidth   Information-carrying capacity of a communication channel. Analog bandwidth is the range of signal frequencies that can be transmitted by a communication channel or network.

BER   *Bit Error Rate.* The number of coding violations detected in a unit of time, usually one second. BER is calculated with this formula: BER=error bits received/total bits sent.

BG   *Border Gateway.* A gate way between the PLMN supporting GPRS and an external inter-PLMN backbone network used to interconnect with other PLMNs also supporting GPRS.

BGP   *Border Gateway Protocol.* A routing protocol that is used to span autonomous systems on the Internet.

Bps   *Bits per second.* Also Kbps or kilo bits per second and Mbps or mega bits per second.

BSC   *Base Station Controller.* The computer controlling a base station the radio equipment.

BSSAP   *Base Station System Application Part.* An interface for procedures between the MSC and the BSS that require interpretation and processing of information related to single calls and resource management, and messages between the MSC and MS which are transparent to the BSC. These messages are handled with SS7 messaging.

BTS   *Base Transceiver Station.* Radio portion of a base station.

CAPEX   *Capital expense.* Expenses that are amortized over time such as major equipment purchases.

Cascading Style Sheets or **CSS**   Cascading style sheets establish style rules that tell a browser how to present a document. One CSS can define the style for an entire Web site.

CCK   *Complementary Code Keying.* A modulation method used in IEEE802.11b to achieve higher data rates and less susceptible to multi-path interference.

CDMA *Code Division Multiple Access.* A method of spread-spectrum communications where all users share the same spectrum at the same time by assigning codes to each user. This offers inherent encryption to the signals.

CDR *Call Detail Record.* An information system that records and reports on telephone calls.

CDPD *Cellular Digital Packet Data.* A digital method of sending data over an analog cellular network.

C-HTML *Compact HTML.* A language used to code content in wireless devices. It is used by the popular i-Mode system. i-Mode is NTT DoCoMo's Internet connection service for mobile phones and is widely used in Japan. Compact HTML is similar to HTML 1.0 and competes with WML.

Circuit Switching Basic switching process whereby a circuit between two users is opened on demand and maintained for their exclusive use for the duration of the transmission.

DARPA *Defense Advanced Research Projects Agency.*

DHCP *Dynamic Host Configuration Protocol.* Software that automatically assigns IP addresses to client stations logging onto a TCP/IP network.

DNS *Domain Name System.* Name resolution software that lets users locate computers on a UNIX network or the Internet (TCP/IP network) by domain name.

DSL *Digital Subscriber Line.* A high-speed digital line for high-speed data access. There are several different versions of DSL including ADSL and HDSL. A DSL is also one channel of an ISDN service.

DSLAM *Digital Subscriber Line Access Multiplexer.* A central office device for ADSL service that intermixes voice traffic and DSL traffic onto a customer's DSL line.

DSSS *Direct Sequence Spread Spectrum.* The data signal is broken up into sequences and transmitted to the receiver, which reassembles the sequences into the data signal.

DTD *Document Type Definition.* A language that describes the contents of an SGML document.

e-Commerce   A term referring to commerce over the Internet. E-commerce is the conduct of monetary transactions via computing device.

EDGE   *Enhanced Data Rate for Global Evolution*. A migration path for TDMA and GSM. GPRS would be added, along with a more robust modulation scheme. Rates to 384 Kbps will offer wireless multimedia IP-based services and applications.

ETSI   *European Telecommunications Standards Institute*.

FHSS   *Frequency Hopping Spread Spectrum*. Frequency hopping spread spectrum continuously changes the center frequency of a conventional carrier several times per second according to a pseudo-random set of channels.

FM   *Frequency Modulation*. A transmission technique that blends the data signal into a carrier by varying the frequency of the carrier.

Gb   Interface between a SGSN and a BSS.

Gc   Interface between a GGSN and a HLR.

Gd   Interface between a SMS-GMSC and a SGSN, and between a SMS-IWMSC and a SGSN.

Gf   Interface between a SGSN and an EIR.

GGSN   *Gateway GPRS Support Node*. Essentially a packet router with some mobility management functions. It connects to the GSM network through standard interfaces.

Gi   Reference point between GPRS and an external packet data network.

GIWU   *GSM Interworking Unit*. An interface to various networks for data communications.

GML   *Generalized Mark-up Language*. A precursor to SGML.

G-MSC or **GMSC**   *Gateway Mobile Station Controller*. An MSC designed to receive wireless calls from the PSTN.

Gn   Interface between two GSNs within the same PLMN.

Gp   Interface between two GSNs in different PLMNs.

GPRS   *General Packet Radio Service*. A packet-based wireless communication service that brings data rates from 56 up to

114 Kbps, and provides continuous connection to the Internet (termed as an "always-on mode") for mobile phone and computer users.

GPS *Global Positioning System.* A satellite-based system for accurate determination of location anywhere in the world.

Gr Interface between a SGSN and a HLR.

Gs Interface between a SGSN and a MSC/VLR.

GSM *Global System for Mobile Communication.* The world standard, started in Europe. Currently it enjoys the largest share of users of any cellular system.

GSN *GPRS Support Node* (xGSN).

GTP *GPRS Tunneling Protocol.* A protocol for security on GPRS.

GW *Gateway.* In interface that performs protocol conversion between different types of networks or applications.

H323 Defines packet standards for terminal, equipment, and services for multimedia communications over local and wide area networks communicating with systems connected to telephony networks such as ISDN.

HDLC *High-Level Data Link Control.* An ISO communications protocol used in X.25 packet switching networks.

HDR *High data rate.* An overlay method compatible with CDMA that allows data rates up to 2.4 Mbps.

Hiperlan2 (Hiperlan) A wireless LAN technology developed by ETSI that provides a 23.5 Mbps data rate in the 5GHz band similar to Ethernet but includes QoS.

HLR *Home Location Register.* A database in a cellular system that contains all the subscribers within the providers home service area.

HSCSD *High-Speed Circuit Switched Data.* A method of increasing data throughput in a circuit-switch connection by using multiple time slots in TDMA or multiple codes in CDMA.

HTML *HyperText Markup Language.* A tag-based language of elements that perform visual presentation or mark-up operations on text within a document.

HTTP   *HyperText Transfer Protocol.* The set of rules for applications exchanging files (text, graphic images, sound, video, and other multimedia files) on the Internet.

HTTP or s-HTTPS   *Secure Hypertext Transfer Protocol.* An extension of HTTP for authentication and encryption between a Web server and browser.

i-Mode   A version of compact HTML utilized by NTT DoCoMo.

IM   *Instant Messaging.* Conferencing using the keypad or keyboard over the Internet or a wireless device between two or more people that requires all parties be online at the same time.

IP   *Internet Protocol.* An implementation of the network layer of the protocol, which contains a network address and is used to route a message to a different network or subnetwork.

IR   *Infrared.* An invisible band of radiation at the lower end of the electromagnetic spectrum commonly used as a wireless communications medium between two devices.

ISDN   *Integrated Services Digital Network.* A digital subscriber service offering either 144 Kbps, suitable for home or SOHO, or 1.544 Mbps, suitable for enterprise markets.

ISM   *Industrial, Scientific and Medical Application Band.* In the 900 MHz and 2.4 GHz bands. Frequencies vary slightly between the U.S. and Europe.

ISO   *International Standards Institute.*

ISP   *Internet Service Provider.* A company that primarily offers Internet access and services to consumers.

Iu   Interface between Radio Network Controller and the SGSN.

IUR   *Internet Usage Record.*

J2ME   *Java Version 2 Mobile Edition.* A compact version of Sun's Java technology targeted for embedded consumer electronics.

JavaScript or JScript   A scripting language technology used in common Web browsers as a client-side technology and also as a server-side integration tool.

JVM *Java Virtual Machine.* A Java interpreter.

L2TP *Layer 2 Tunneling Protocol.* A protocol from the IETF for creating virtual private networks over the Internet.

LLC *Logical Link Control.* Provides a common interface point to the MAC layers, which specify the access method used.

MAC *Medium Access Control.* The protocol that controls access to the physical transmission medium on a LAN.

m-Commerce Mobile commerce or mobile e-commerce. Mobile commerce is the conduct of monetary transactions via a mobile device such as a WAP device.

MGCP/Megaco *Media Gateway Control Protocol.* Specifies communication between call control elements and telephony gateways in VoIP networks. Megaco was created for the same purpose, but it differs from MGCP because it supports a broader range of networks and devices.

MILNET *Military Network.*

MM *Mobility management.* The process of assigning and controlling of wireless links for terminal network connections of cellular devices.

MMS *Multimedia Messaging.* Pager and SMS messaging which includes graphics or video components.

MPLS *Multi-Protocol Label Switching Router.* A specification for layer 3 switching from the IETF.

MS *Mobile station.* A cellular or PCS radio.

MSC *Mobile Switching Center.* A switching center for mobile customers that connects them to each other or the public switched network.

MVNO *Mobile Virtual Network Operators.* A mobile operator that does not own its own spectrum and usually does not have its own network infrastructure that establishes business arrangements with traditional mobile operators to buy minutes of use for sale to their own customers.

NAS *Network access server.* A specialized file server that connects to the network.

NMT *Nordic Mobile Telephone System.* An early cellular system in Scandinavia and Europe.

NSF *National Science Foundation.*

OA&M *Operations, administration, and management.*

OFDM *Orthogonal Frequency Division Multiplex.* A method of digital modulation in which a signal is split into several narrowband channels at different frequencies.

OPEX *Operation expense.* The expenses of daily operations.

OSI Seven-Layer Model A standard architecture for data communications. Layers define hardware and software required for multivendor information processing equipment to be mutually compatible. The seven layers from lowest to highest are: Physical, Link, Network, Transport, Session, Presentation, and Application.

OSS *Operations Support System.* A support system for deploying new IP-based services (VPN, QoS, VoIP) which enables rapid service provisioning and service level agreement management.

PCU *Packet control unit.*

Packet Switching or **PS** An efficient method for breaking down and handling high-volume traffic in a network. A transmission technique that segments and routes information into discrete units. Packet switching allows for efficient sharing of network resources because packets from different sources can all be sent over the same channel in the same bitstream.

PDA *Personal Digital Assistant.* A hand-held device such as the Palm Pilot.

PDC *Personal Digital Cellular.* A Japanese standard very similar to TDMA in the U.S.

PDN *Packet Data Network.* An IP network for packet data.

PLMN *Public Land Mobile Network.* Generic name for all mobile wireless networks that use earth base stations rather than satellites; the mobile equivalent of the PSTN

PSPDN *Packet Switched Public Data Network.* A public packet data network such as the Internet.

PSTN *Public Switched Telephone Network.* The public telephone circuit switched network.

PVC *Permanent virtual circuit.* A point-to-point connection that is established ahead of time.

QoS *Quality of service.* The ability to define a level of performance and priorities in a data communications system.

RADIUS *Remote Authentication Dial-In User Service.*

RAN *Radio Access Network.* The radio access network provides the basic transmission, radio control, and management functions needed for the mobile subscriber to access the resources of the core network and the end-user services network.

RFC *Request for Comment.* A mechanism for development of Internet procedures and specifications; RFCs create a "standard" without the formal ratification process (such as an ANSI standard).

RLP *Radio Link Protocol.* A handshaking protocol for handling lost data in wireless transmissions.

RNC *Radio Network Controller.* The intelligence of an RAN.

SCP *Service Control Point.* A node in an SS7 telephone network that provides an interface to databases, which may reside within the SCP computer or in other computers.

SGML *Standardized General Mark-up Language.* An ISO standard for defining the format in a text document.

SGSN *Serving GPRS Support Node.* The node is responsible for handling packet data to and from a mobile device.

SIP *Session Initiation Protocol.* An Application layer control protocol that can establish, modify, and terminate multimedia sessions or calls.

SLA *Service-level agreement.* A contract between the provider and the user that specifies the level of service that is expected during its term.

SMS *Short Message Service.* A messaging service supported by many mobile phones that allows short text messages, typically in the range of 120 characters, to be sent between mobile devices.

SMSC   *Short Message Service Center.* A temporary message storage facility for flow control,

SS7   *Signaling System Number 7.* The protocol used in the public switched telephone system for setting up calls and providing services.

SOHO   *Small Office Home Office.* A market of small businesses, some operating out of a home environment.

TACS   *Total Access Communications System* (replaced by enhanced version—ETACS).

TCP/IP   TCP (Transmission Control Protocol) is a method (or protocol) used in conjunction with the Internet Protocol (IP) to send data in the form of message units (datagrams or packets) between computers over the Internet. IP handles the actual delivery of the data and TCP tracks the individual data units so that the message can be reconstructed on the receiving end.

TDMA   *Time Division Multiple Access.* A method of dividing a communications channel into time slots with different users assigned to each slot. In the United States, this is known as *IS-136.*

TE   *Terminal equipment.*

TIA   *Telecommunications Industry Association.*

TS   *Time slot.*

Um   Interface between the MS and the GPRS fixed network part.

UMTS   *Universal Mobile Telephone System.* The name of a new mobile networking standard that will supplement and ultimately replace GSM. UMTS is considered a Third Generation Cellular System using higher data rates than currently possible.

UNII   *Unlicensed National Information Infrastructure Band*—5.15 GHz to 5.85 GHz.

URL   *Uniform Resource Locator.* Addresses of Web-based resources; they can refer to static pages and to applications (scripts).

UTRAN   *UMTS Terrestrial Radio Access Networks.*

VAS *Value-added Services.*

VHE *Virtual Home Environment.* A concept that a network supporting mobile users should provide them the same computing environment on the road that they have in their home or corporate computing environment.

VLR *Visitor Location Register.* A database in a cellular system that contains all the subscribers that are currently visiting within this service area.

V-MSC *Voice Mobile Station Controller.* The MSC for the circuit switched voice traffic.

VPN *Virtual Private Network.* A private network that is configured within a public network.

VoIP *Voice-Over-Internet Protocol.* A method of breaking voice communications into packets and routing them just like any other Internet traffic.

W3C *World Wide Web Consortium.*

WAP *Wireless Application Protocol.* A specification for a set of communication protocols to standardize the way that wireless devices, such as cellular mobile telephones, can be used for Internet-based access. WAP's protocol layers are as follows:

- Wireless Application Environment (WAE)
- Wireless Session Layer (WSL)
- Wireless Transport Layer Security (WTLS)
- Wireless Transport Layer (WTP)

W-CDMA *Wideband Code Division Multiple Access.* A form of CDMA using higher bandwidth by utilizing wider radio frequency spectrum.

WID *Wireless Information Device.*

WLAN *Wireless Local Area Network.* A wireless implementation of a local area network such as a wireless Ethernet or HomeRF network.

WML *Wireless Markup Language.* A tag language that allows the text portions of Web pages to be presented on cellular phones or Personal Digital Assistants via wireless access.

WML is similar in appearance to HTML. An alternative to WML is compact HTML.

WMLScript   A scripting language for use with WAP devices, similar to JavaScript.

WTAI   *Wireless Telephony Application Interface.* The WTAI specification describes standard telephony-specific extensions for WAP devices with WML and WMLScript interfaces to such items as call control features, address book, and phonebook services.

XHTML   *Extensible HTML.*

XML   *Extensible Markup Language.* The World Wide Web Consortium's standard for Internet Markup Languages. WML is one such language. XML describes the *structure* of content, unlike HTML, which describes how pages appear when viewed.

INDEX

Note: Boldface numbers indicate illustrations; italic *t* indicates a table.

ABOUT THE AUTHORS

Roman Kikta is the General Partner and co-founder of Genesis Campus, a technology business creation incubator and venture fund. Mr. Kikta has worked for leading wireless industry companies: Nokia Mobile Phones, Panasonic Communications and Systems Company, GoldStar, and OKI Telecom. Mr. Kikta is an acknowledged expert on market trends globally from anthropological, sociological,

and psychological perspectives and on the role of technologies on society. He has led development of several industry firsts including: cellular payphone, cellular in-building system/PBX adjunct, and voice recognition dialer, in addition to several generations of mobile, transportable, and portable phone designs, features, and functions, as well as the initial product launch of PCS in the US. Mr. Roman Kikta has also co-authored other books published by McGraw-Hill on communications technologies, including *Delivering xDSL* and *3G Wireless Demystified*.

Al Fisher is a Partner of Genesis Campus, a venture capital firm. Mr. Fisher is a wireless telecommunications specialist with over twenty years in engineering, manufacturing, and marketing with leading telecom equipment manufacturers,

including OKI Telecom, Uniden America, and Anritsu Company. He was a co-founder and V.P. of Engineering for NET-TEL Technologies, Inc., a Research & Development firm specializing in biometric security with wireless devices and computer networks and served as President of Concept Technologies, a design consulting firm. Mr. Fisher has served on several TIA and ANSI committees, chairing several subcommittees, and served as a member of the Advisory Committee for Penton Publishing technical programs on Wireless Internet. He has presented several technical seminars in the United States and internationally, and also authored several articles in technical journals. Mr. Fisher is a co-holder for patents in secured access using biometrics and an interactive messaging and programming system using in-band signaling and voice recognition technology. Mr. Fisher has received his BS degree from Drexel University.

Michael Courtney is a Partner with Genesis Campus, a technology business creation incubator and venture fund. Mr. Courtney has over twelve years experience in Marketing Research and Business Development in financial and telecommunications industries. Mike has conducted extensive market research to test new product and service concepts, demand, positioning, and pricing for companies  such as AT&T, Lucent, Nortel, Ericsson, and Nokia. Mr. Courtney has extensive knowledge of telecommunications infrastructure, wireless devices, and emerging technologies. Mr. Courtney has been involved with product and concept development of technologies including wireless instant messaging, Bluetooth, Java, WAP, Biometrics, and mobile payment systems. He is a frequent seminar leader and speaker at corporate, industry and academic events. Mr. Courtney has a B.S. in Business Administration from State University of New York and a Masters of Marketing Research from the University of Georgia, Athens, GA.

Wireless Crash Course
Paul Bedell
0-07-137210-5 / 2001 / 400 pgs

How Wireless Works in the Real World

In this much-needed resource, Bedell, who has built out fixed, interconnection, and WAN networks for three major wireless carriers, leaves heavy-duty math to the scientists and gives you wireless in plain English. Step-by-step, he shows you how wireless voice and data systems work.

Packed with easy-to-understand information, *Wireless Crash Course* is the perfect text for a first course in wireless management—and an ideal tool for managers and engineers who want to expand their understanding of how wireless carrier systems work, and how to make them work better at a fraction of the cost.

One Book that Explains It All

- Cellular Radio Systems
- Cellular Regulatory Structure
- Fundamental Design Parameters
- Cellular System Components and Design
- Criteria and Methods of Cell Placement
- AMPS Tech Specifications
- RF Channelization, Propagation and Power
- Towers • Antenna Types and Uses
- LMDS v MMDS • Bluetooth
- WAP • 3G • Paging • EMSR • Satellite-Based PCS
- Cell Site Equipment and RF Signal Flow
- Cellular System Capacity Engineering • Regulatory Processes
- Enhancers • Microcells • Tools and Testing
- Mobile Telephone Switching Office
- N-AMPS Cellular Standard • COWs
- Fixed Network and System Connectivity
- Interconnection to the PSTN • Digital Cellular Systems vs PCS
- Cellular Call Processing • Network Operations Centers
- Intercarrier Networking • Wireless Fraud